Numerical Reasoning Testing Guide

www.How2Become.com

Orders: Please contact How2become Ltd, Suite 2, 50 Churchill Square Business Centre, Kings Hill, Kent ME19 4YU.

You can order through Amazon.co.uk under ISBN 9781910602027, via the website www.How2Become.com or through Gardners.com.

ISBN: 9781910602027

First published in 2015 by How2become Ltd.

Typeset for How2become Ltd by Anton Pshinka.

Printed in Great Britain for How2become Ltd by:
CMP (uk) Limited, Poole, Dorset.

CONTENTS

INTRODUCTION
TO YOUR NEW GUIDE

INTRODUCTION TO YOUR NEW GUIDE

Welcome to your new guide, Numerical Reasoning. This guide is a comprehensive testing book which provides lots of practice questions for **basic, intermediate** and **advanced** mathematics.

This guide contains a variety of mathematical questions for anyone who is asked to take a Numerical Reasoning test.

The key to success for psychometric testing is through practice and preparation. We have provided you with lots of questions in order for you to gain a complete understanding of what you are likely to face in a Numerical Reasoning test.

It is important that when working through this book, you check your answers at the end of each testing chapter. Knowing where you went wrong, and understanding what you need to do to reach the correct answer, is just as important as getting the correct answer. If you know how to fix your mistakes, you are far more likely to get the answer correct next time.

Good luck and we wish you all the best.

The how2become team

The How2Become Team

STRUCTURE OF THE BOOK

In order to make the most out of your new guide, it is important to understand the structure of the testing book. We have done our utmost to create a guide that is suitable for all mathematical abilities; to assist you in passing any Numerical Reasoning Test.

Your Numerical Reasoning guide has been broken down into three main sections: basic, intermediate, and advanced. This is to ensure your mathematical skills are tested in the best way possible.

This comprehensive Numerical Reasoning guide follows the structure as formulated below:

- Introduction to your New Guide
    - What is Numerical Reasoning?
    - Aims and Objectives
    - What to Expect
    - Numerical Reasoning Testing Strategies
- Numerical Reasoning – **Basic**
    - Detailed Example Questions
    - Numerical Reasoning – Basic (Section 1) and (Section 2)
    - Detailed Answers and Explanations
- Numerical Reasoning – **Intermediate**
    - Detailed Example Questions
    - Numerical Reasoning – Intermediate (Section 1) and (Section 2)
    - Detailed Answers and Explanations
- Numerical Reasoning – **Advanced**
    - Detailed Example Questions
    - Numerical Reasoning – Advanced (Section 1) and (Section 2)
    - Detailed Answers and Explanations
- A Few Final Words...

Whilst we do not provide an exact account of what your Numerical Reasoning test will look like, we do provide an insight into what you can expect in terms of questions. The most important thing you need to understand before sitting any Numerical Reasoning test, is how to answer the questions.

Fundamentally, our How2become team have designed this guide to:

- Ensure you are fully prepared for a Numerical Reasoning test.
- Provide different levels of difficulty: **basic**, **intermediate** and **advanced**.
- Provide detailed answers and explanations for you to fully comprehend how to reach the correct answer.
- Demonstrate sample questions, providing a step-by-step account of what you need to do in that particular arithmetic.

WHO TAKES A NUMERICAL REASONING TEST?

A Numerical Reasoning test is often used in job selection processes to determine whether or not you are suitable for the job role. Numerical Reasoning tests are a useful tool for employers in terms of screening applicants, and are often used to narrow the interview field.

This test will measure your ability to solve mathematical problems and equations. Through practice and perseverance, we will ensure that by the end of the book, you have the knowledge and understanding to answer questions at an array of difficulty levels.

WHAT ARE NUMERICAL REASONING TESTS?

A Numerical Reasoning test is designed to assess mathematical knowledge through number-related assessments. These assessments can be of different difficulty levels, and will all vary depending on who you are sitting the test for. So, be sure to find out what type of Numerical Reasoning test you will be sitting, to ensure you are fully able to practice prior to the assessment.

The majority of Numerical Reasoning tests are administered to candidates who are applying for managerial, graduate and professional positions; any job that deals with making inferences in relation to statistical, financial or numerical data. However, some employers may use these tests as a way of determining important job-related skills such as time management and problem solving efficiency.

Numerical Reasoning tests cover a wide range of mathematical formulas; and so it is imperative to comprehend the skills and knowledge required to work out the mathematics involved. Most Numerical Reasoning tests contain questions in relation to:

Adding	Subtracting	Dividing	Multiplying
Fractions	Percentages	Decimals	Ratios
Charts and Graphs	Mean, Mode, Median, Range	Areas and Perimeters	Number Sequences
Time	Conversions	Measurements	Money
Proportions	Formulae	Data Interpretation	Quantitative Data
Data Analysis	Correlations	Statistics	Shapes

WHAT SKILLS ARE MEASURED?

Obviously, a Numerical Reasoning test primarily deals with assessing your level of mathematical ability. Other skills that are also measured, and often assessed by examiners are:

- Critical Reasoning
- Estimations
- Speed
- Concentration
- Analysis
- Interpretation

PREPARING FOR A NUMERICAL REASONING TEST

Your performance in a Numerical Reasoning test can undoubtedly be bettered through practice! Getting to grips with the format of the test, and gaining an insight of the typical questions you are likely to face will only work to your advantage.

The more you practice, the more you will see your performance excel! With any psychometric testing, it is important to fully maximise your skills and knowledge prior to your assessment to ensure the best result.

This comprehensive guide will provide you with lots of sample questions, similar to those that will be found on your Numerical Reasoning test. Our insightful and ultimate preparation guide will allow you to grasp each question type, understand what is expected, and show you how to achieve the correct answer.

WHAT TYPES OF NUMERICAL TESTS ARE THERE?

Numerical Reasoning tests vary in their format, in both types of question and difficulty. However, they all test similar arithmetic. Before taking your actual test, we advise that you research the type of test that you will be required to sit. The key components, which distinguish different Numerical Reasoning tests; are **Format** and **Difficulty**.

Formats:

- **Graph** and **Charts** – to interpret and analyse data and answer the following questions in relation to that data.
- **Word problems** – short word problems or passages that deal with riddles and/or calculations.
- **Number Sequences** – the ability to find the pattern or correlation amongst a sequence.
- **Basic math** – demonstrate basic arithmetical understanding.

Level of Difficulty:

- **Basic** – Basic GCSE Maths - simple mathematical formulas and calculations, interpretation, analysis.
- **Intermediate** - Strong GCSE Maths – interpretation, equations, charts and graphs, statistics.
- **Advanced** – Higher Level Maths - critical reasoning and analysis, quantitative reasoning.

Levels of difficulty and different formats are determined by the job for which you are applying. Your test will solely depend on the nature of the job and the position you are applying for, and therefore the requirements for each test and desired level of ability, will vary.

NUMERICAL REASONING STRATEGIES

Helpful Tips

- Questions will often require you to identify what mathematical formulae is being used (division, percentage, ratio etc). Before you answer the question, carefully read what the question is asking you to do! Be sure to understand what you need to work out, before attempting to answer the question.
- Do not spend too much time on one particular question. You may find some questions easier than others. You may struggle at a certain 'type' of question and so it is important not to ponder over questions you are unsure of. Move on and then come back to those questions at the end.
- Accuracy is key; avoid silly mistakes! You need to remain as accurate as possible to ensure a high and successful score. That's why it is important to fully comprehend the questions and understand what is being asked.
- These tests are designed under strict time limits. Psychometric testing is fundamentally used to measure people's level of accuracy, whilst working in speedy conditions.
- Working out mental arithmetic can be difficult. Do not be afraid to write down your calculations.
- Practice is key. The more you practice your mental arithmetic and other mathematical formulae; the easier it becomes. This is why we have provided you with lots of sample questions for you to work through. The more you practice these tests, the more likely you are to feel comfortable and confident with the questions. Remember, practice makes perfect!
- If you are unsure about the answers, make sure you use our detailed answers and explanations to understand how to reach the correct answer. Remember, knowing where you went wrong is just as important as getting the questions correct. Try practising the question again after reading through the answers and explanations to ensure you know where you went wrong.
- Our guide is broken down into three main sections: basic, intermediate and advanced. If you find one testing section relatively easy, maybe try the next level of difficulty. The more you test yourself and your ability; the more confident you will feel when it comes to tackling a numerical test – no matter what level of difficulty it is!

TIPS FOR PASSING NUMERICAL REASONING

Practice!

- No great accomplishment comes easy! You have to work hard at it! Perseverance and practice are two important things to remember when sitting a Numerical Reasoning test. Nothing will boost your chances at success more than if you practice them prior to your assessment. Not only will this provide clarity and understanding of what to expect, but it will also take off some of the pressure you may be feeling before that all important test!

Stay calm.

- If you lose focus or become overwhelmed during your Numerical Reasoning test, it is highly likely that this will impact your overall performance. Try to stay calm and focused throughout your assessment. Remember, if you practice prior to the test, you will have far more experience and knowledge going into the test itself, and this is invaluable when comparing your results with others who did not practice, or had no knowledge of the paper beforehand.

You can always work backwards!

- If you get stuck, why not try the sequence in reverse? This will allow you to visualise the sequence from a different perspective; and allow you to spot something you may have missed previously.

Manage your time.

- The time you are allowed to complete your Numerical Reasoning test will very much depend on your circumstances. Try to find out how long your test is going to last, and use this information to your advantage. Managing your time in psychometric tests is significant; practising these tests prior to your assessment will give you some indication of how well you will perform under extreme time restrictions.

Finally, we have also provided you with some additional free online psychometric tests which will help to further improve your competence in this particular testing area. To gain access, simply go to:

www.PsychometricTestsOnline.co.uk

Good luck and best wishes,

The how2become team

The How2become team

Numerical Reasoning – ***BASIC***

Our Numerical Reasoning **(basic)** section will provide you with the skills and knowledge expected for basic GCSE level mathematics. The difficulty of the questions will depend on the type of Numerical Reasoning test that you are going to be taking.

In order for you to successfully pass a Numerical Reasoning test, we have done our utmost to ensure that you have as many questions as possible, that focus specifically on the basics.

In this type of basic Numerical Reasoning test, you can expect to find questions on the following areas:

- Percentages
- Fractions
- Decimals
- Areas
- Perimeters
- Angles
- Symmetry
- Inputs and Outputs
- Data Interpretation
- Prime, Multiple and Factor numbers
- Mean / Mode / Median / Range

Whilst we have provided you with an array of questions, your Numerical Reasoning test will be tailored more specifically to the job for which you are applying, and so the questions may not be the same, but will ultimately test the same skills and knowledge in terms of basic arithmetic.

EXAMPLES of Numerical Reasoning - ***BASIC***

Adding Fractions

$$\frac{5}{7} + \frac{3}{5}$$

$$\frac{5}{7} \times \frac{3}{5} = \frac{25 + 21}{35} = \frac{46}{35} = 1\frac{11}{35}$$

Crossbow Method:

The CROSS looks like a multiplication sign and it tells you which numbers to multiply together.

One arm is saying 'multiply the 5 by the 5', and the other arm is saying 'multiply the 7 by the 3'.

The BOW says 'multiply the 2 numbers I am pointing at'. That is 7 times 5.

The answer is 35 and it goes **underneath** the line in the answer.

Subtracting Fractions

$$\frac{4}{7} - \frac{2}{5}$$

$$\frac{4}{7} \times \frac{2}{5} = \frac{20 - 14}{35} = \frac{6}{35}$$

To subtract fractions, the method is exactly the same. The only difference is, you minus the two numbers forming the top of the fraction, as opposed to adding them.

Multiplying Fractions

$$\frac{2}{3} \times \frac{4}{7}$$

$$\frac{2}{3} \times \frac{4}{7} = \frac{8}{21}$$

Arrow Method:

Multiplying fractions is easy. Draw two arrows through the two top numbers and the two bottom numbers (as shown above) and then multiply – simple!

Sometimes the fraction can be simplified, but in the above example, the answer is already in its simplest form.

Drawing the arrows is just to help you remember which numbers to add up. Once you have mastered this knowledge, try doing it without the arrows.

Dividing Fractions

$$\frac{3}{7} \div \frac{1}{3}$$

$$\frac{3}{7} \div \frac{3}{1} = \frac{3}{7} \times \frac{3}{1} = \frac{9}{7} = 1\frac{2}{7}$$

Most people think that dividing fractions is difficult. However, it's actually relatively simple if you have mastered multiplying fractions.

Mathematicians realised that if you turn the second fraction upside down (like in the above example), and then change the 'divide' sum to a 'multiply', you will get the correct answer.

Simplifying Fractions

$$\frac{24}{30} = \frac{12}{15} = \frac{4}{5}$$

Simplifying Fractions

There are a few steps to follow in order to correctly simplify fractions.

- Can both numbers be divided by 2? If yes, then how many times does 2 go into each number? Write the new fraction.
- Using the new fraction, do the same thing. Can 2 go into both numbers? If yes, divide both numbers by 2.
- If both numbers cannot be divided by 2, then try the first odd number: 3. Can both numbers be divided by 3? If yes, divide both numbers by 3. Do this again until 3 no longer goes into the number.
- If 3 does not go into the numbers again, it doesn't mean it's finished. Try the next odd number: 5, and so on until the fraction can no longer be simplified.

Fractions and Numbers

What is $\frac{3}{7}$ of 700?

How to work it out:

- $700 \div 7 \times 3 = 300$.

Percentages

What is 45% of 500?

How to work it out

- To work out percentages, divide the whole number by 100 and then multiply the percentage you want to find.
- **For example:**
  - o 500 ÷ 100 x 45 = 225
  - o So, 225 is 45% of 500.

Fractions / Decimals / Percentages

$$\frac{1}{10} = 0.1 = 10\%$$

How to turn fractions into decimals, and decimals into percentages

- 0.1 into a percent, you would move the decimal point two places to the right, so it becomes 10%.
- To convert 1/10 into a decimal, you would divide both numbers. For example, 1 ÷ 10 = 0.1.
- To convert 10% into a decimal, you move the decimal point two places to the left. For example, to convert 10% into a decimal, the decimal point moves two spaces to the left to become 0.1.

Volume

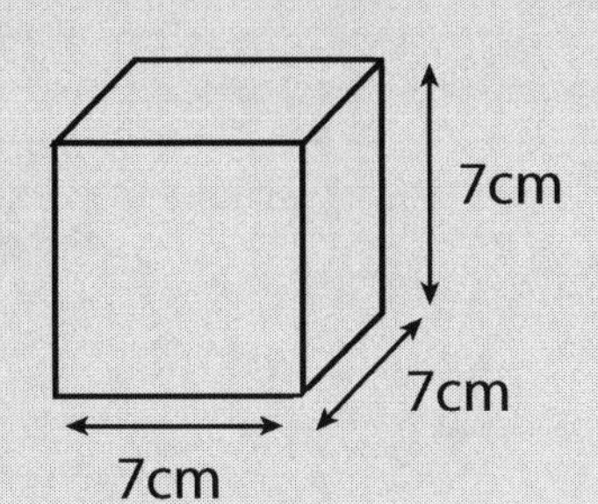

Volume

Length x base x height

- **7 x 7 x 7 = 343**

Areas / Perimeters

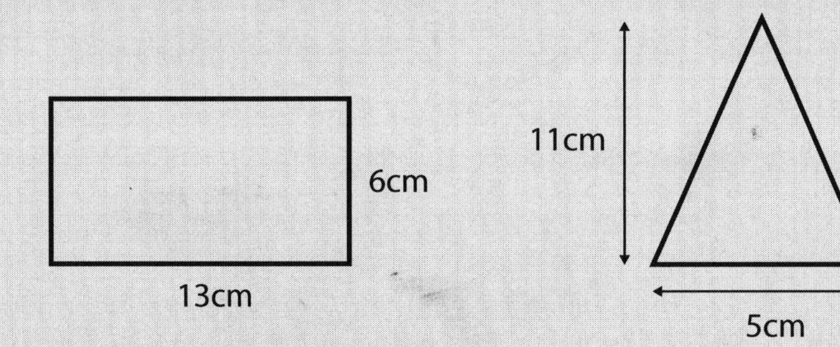

Area of squares/ rectangles

Base x height

- 13 × 6 = 78 cm²

Area of triangles

½ base x height

- 11 × 5 ÷ 2 = 27.5

Perimeter

Add all the sizes of each side.

- 6 + 6 + 13 + 13 = 38

Angles

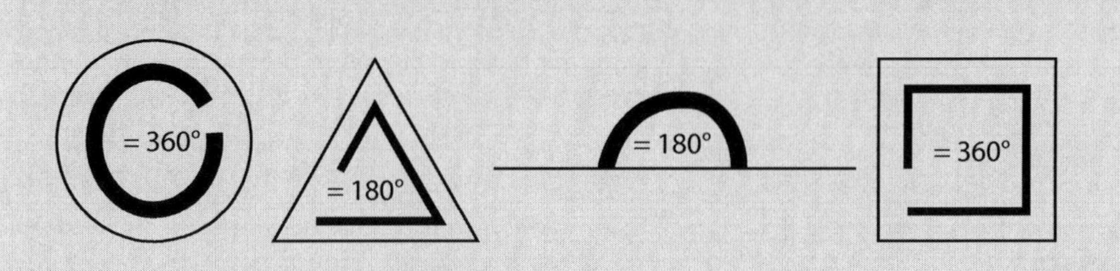

Symmetry

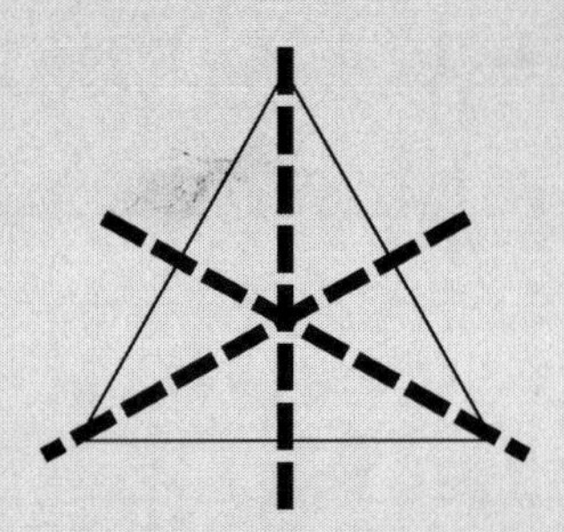

How to work it out:

- To work out how many lines of symmetry a shape has, you need to see where the shape can be folded, in order to create the same reflection.
- Note, an equilateral triangle has 3 lines of symmetry because it can be rotated 3 turns. The triangle would look exactly the same for each rotation.
- **Remember, don't count the same line of symmetry more than once!**

Inputs and Outputs

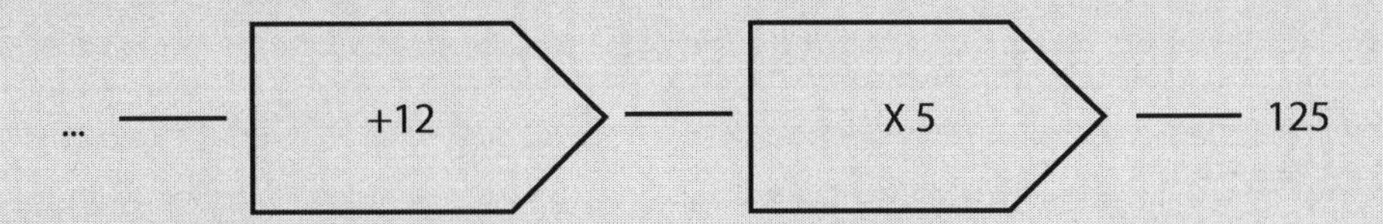

How to work it out:

In order to work out the missing number at the start of the sequence, you will need to work backwards.

- When working backwards, you need to do the OPPOSITE.
- For example:
  - o 125 ÷ 5 – 12 =13
- You can factor '13' into the equation to make sure you have the correct answer, and the equation works.

Simplifying Equations

Simplify 5w - 6x - 2w - 1x

(5w) (- 6x) (- 2w) (- 1x)

(5w - 2w) = 3w

(-6x - 1x) = -5x

3w - 5x

- The important thing to remember for simplifying equations is to break up the equation (like above).
- The '-' signs and the '+' signs should also be grouped and be on the left side of the number.

Number Sequences

13, 26, 52, 104, 208, 416, ... , ...

How to work it out:

In order to work out number sequences, you need to understand what is happening from one number to the next.

- For example, in the above number sequence: you should be able to see the pattern of 'doubling'.
- So, the next number after 416 should be double: 832.
- The next number after 832 should be double: 1664.

Ratios

Ben has some sweets. He is going to share them with his two friends. Ben has 24 sweets and is going to share them in the ratio of 4 : 2 : 2.

How many sweets does each person get?

- Add up the ratios = 4 + 2 + 2 = 8.
- 24 ÷ 8 = 3.
- So, 3 x 4 = 12.
- 3 x 2 = 6.
- 3 x 2 = 6.

So one person will have 12 sweets and the two other people will get 6 sweets.

Prime Numbers

2	3	5	7	11	13	17	19
23	29	31	37	41	43	47	53
59	61	67	71	73	79	83	89

A prime number is a number that can only be divided by 1 and itself.

- For example, no other numbers apart from 1 and 5 will go into 5.

Factors

Factors are numbers that can be divided into the original number. For example, 6 has the factors of 1 and 6, 2 and 3.

Factors of 12:

Factors are all the numbers that can go into the number.

So, $1 \times 12 = 12$
2×6
3×4

So in ascending order, 1, 2, 3, 4, 6 and 12 are all factors of the number 12.

Multiples

- A multiple is a number which is made from multiplying a number in the same pattern.
- For example, the multiples of 2 are: 2, 4, 6, 8, 10, 12, 14 etc.
- Multiples of 15 are: 15, 30, 45, 60, 75 etc.

Speed / Distance / Time

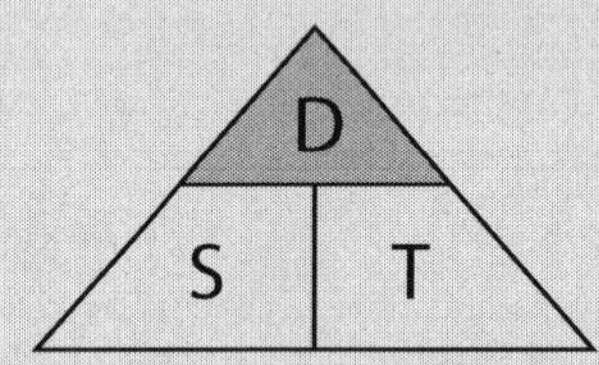

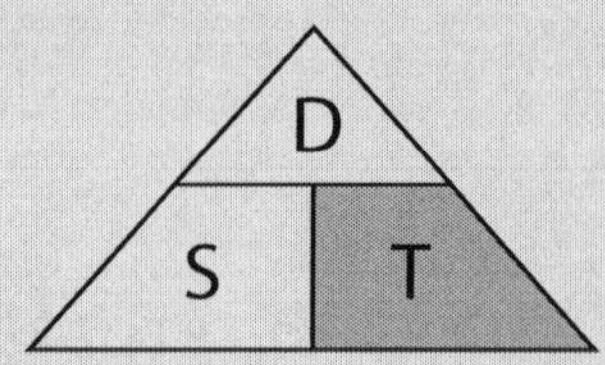

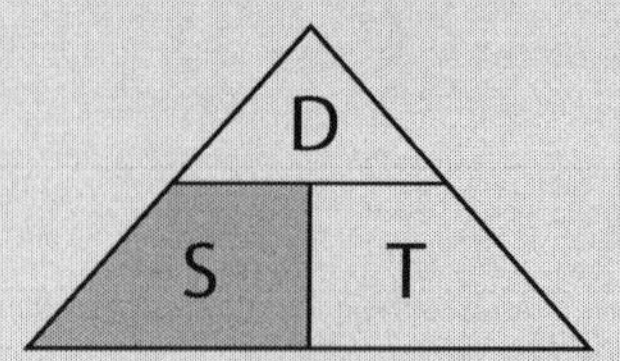

To work out the Distance:

- Distance = Speed x Time

To work out the Time:

- Time = Distance ÷ Speed

To work out the Speed:

- Speed = Distance ÷ Time

Tenths / Hundredths /Thousandths

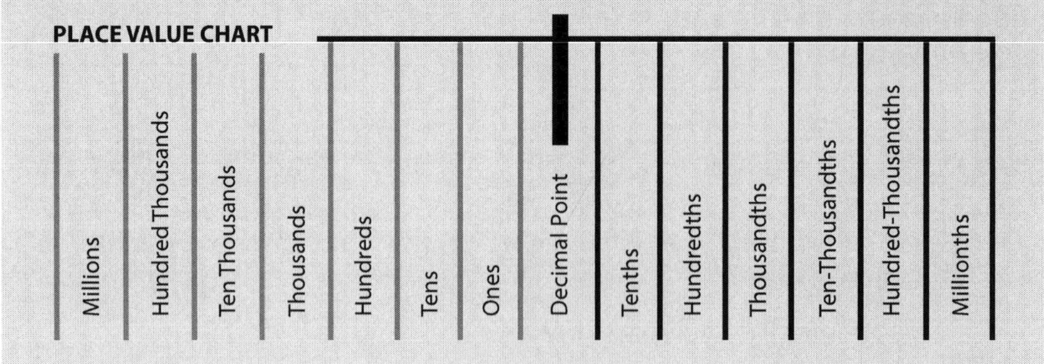

Stem and Leaf Diagrams

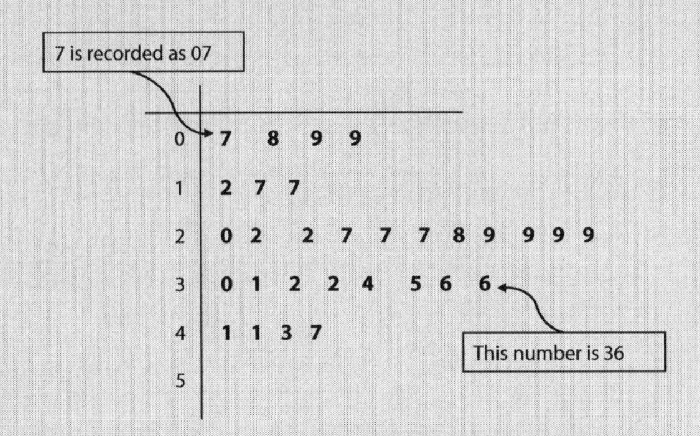

- Stem and leaf diagrams act as a way of handling data.
- These become particularly useful when dealing with large sums of data.
- They are also helpful ways to work out the **mean**, **mode**, **median** and **range**.

Mean / Mode / Median/ Range

Mean

- To work out the mean of a set of data, you add up all the numbers and then divide the total value by the total amount of numbers.

Mode

- The mode is easily remembered by referring to it as the 'most'. What number occurs most throughout the data?

Median

- Once the data is in ascending order, you can then work out what number is the median. In other words, what number is in the middle? If no number is in the middle, use the two numbers that are both in the middle; add them up and divide by 2.

Range

- In ascending order, the range is from the smallest number to the biggest number.

*Numerical Reasoning – **BASIC***

(Section 1)

Question 1

A charity arranges a bike race. 120 people take part. 1/3 of the people finish the race in under half an hour. How many people did not finish the race in under half an hour?

Answer

Question 2

What is 3/5 of 700?

Answer

Question 3

There are 4000 millilitres of water contained in the jug. If 1 litre is equivalent to 1000 millilitres, how many litres of water are there?

Answer

Question 4

What is the missing angle?

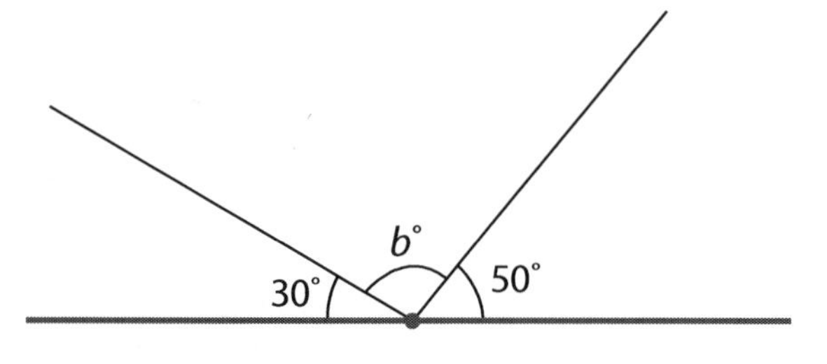

Answer

Question 5

What is 120 multiplied by 13?

Answer

Question 6

Find 60% of £45.

Answer

Question 7

How many lines of symmetry does this shape have?

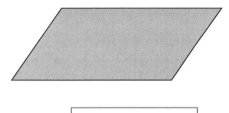

Answer

Question 8

A packet of biscuits weighs 120 g. Find the weight of 9 packets of biscuits.

A	B	C	D
1080 kg	1880 g	1080 g	108 kg

Question 9

A squared field has a perimeter of 72 cm. What is the area of the squared field?

Answer

Question 10

What is 24/48 in its simplest form?

Question 11

Look carefully for the pattern, and then choose which pair of numbers comes next.

5 7 9 11 13 15 17 … …

A	B	C	D
18, 19	19, 21	19, 20	21, 23

Question 12

Look carefully for the pattern, and then choose which pair of numbers comes next.

0 1 1 2 3 5 8 … …

A	B	C	D
12, 18	13, 21	15, 23	13, 22

Question 13

Liz has £12.00. Steph has £8.50.

What is the ratio of Liz's money to Steph's money, in its simplest form?

Answer

Question 14

A newspaper includes 16 pages of sport and 8 pages of TV. What is the ratio of sport to TV? Give your answer in its simplest form.

Answer

Question 15

Multiply 6 by 7 and then divide by 3.

Answer

Question 16

Divide 120 by 4 and then multiply it by 5.

Answer

Question 17

What is 9/11 of 88?

Answer

Question 18

An English class of 28 have just sat a mock Exam. The exam has 2 sections – Literature and Language. It takes approximately 6 minutes to mark the Literature section and 7 minutes to mark the Language section. Another 2 minutes is given on each exam to check the work again. How long in hours and minutes does it take to mark the English mock exam?

A	B	C	D
6 hours and 45 minutes	5 hours and 25 minutes	7 hours	9 hours and 10 minutes

Question 19

What is 0.9 as a percentage?

A	B	C	D
0.009%	0.9%	9%	90%

Question 20

Simplify x + 8x – 3x.

A	B	C	D
5x	6x	7x	12x

Question 21

Work out 23.7 – 2.5 × 8.

Answer [ ]

Question 22

There are 20 buttons in a bag. 12 are red, 5 are green and the rest are white. A button is chosen at random. Work out the probability that it is white.

Answer [ ]

Question 23

On a school trip at least 1 teacher is needed for every 8 students. Work out the minimum number of teachers needed for 138 students.

Answer

Question 24

Translate the triangle so that point A moves to point B. Draw your translation on the graph.

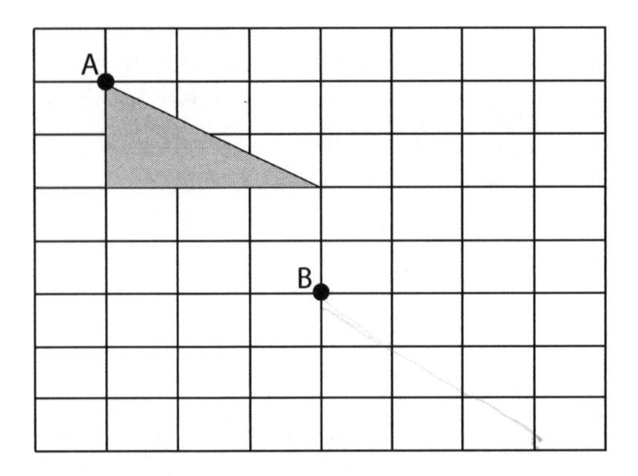

Question 25

Rotate the triangle 90° clockwise so that point A moves to point B. Draw your rotation on the graph.

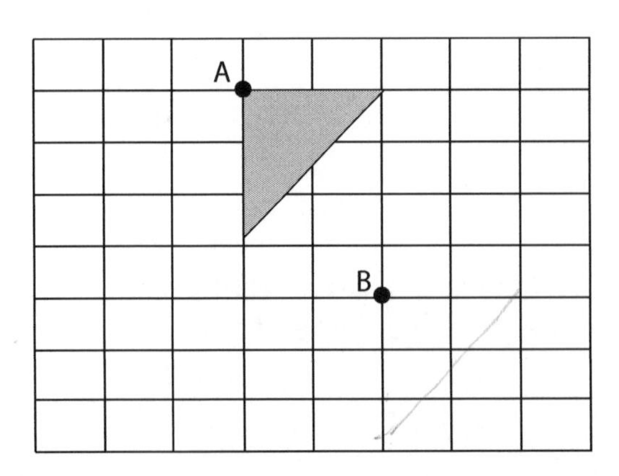

Question 26

The school day starts at 0845. They have 15 minutes form time and then a 25 minute assembly before the first lesson starts. What time does the first lesson start?

Answer

Question 27

A cinema has 27 rows of seats, 28 seats in each row. Tickets are £8 each.

The cinema has sold tickets for every seat apart from 5. Estimate how much, to the nearest hundred, the cinema will make, based on the information provided?

Answer

Question 28

How many grams are there in 2.5 kilograms?

A	B	C	D
0.0025g	250g	2005g	2500g

Question 29

What is the value of 9 in 5.92?

A	B	C	D
9/10	1/9	1/90	9/100

Question 30

The scatter graph shows the number of driving lessons and the number of tests taken to pass by 10 people.

What proportion of the 10 people, passed on their first test?

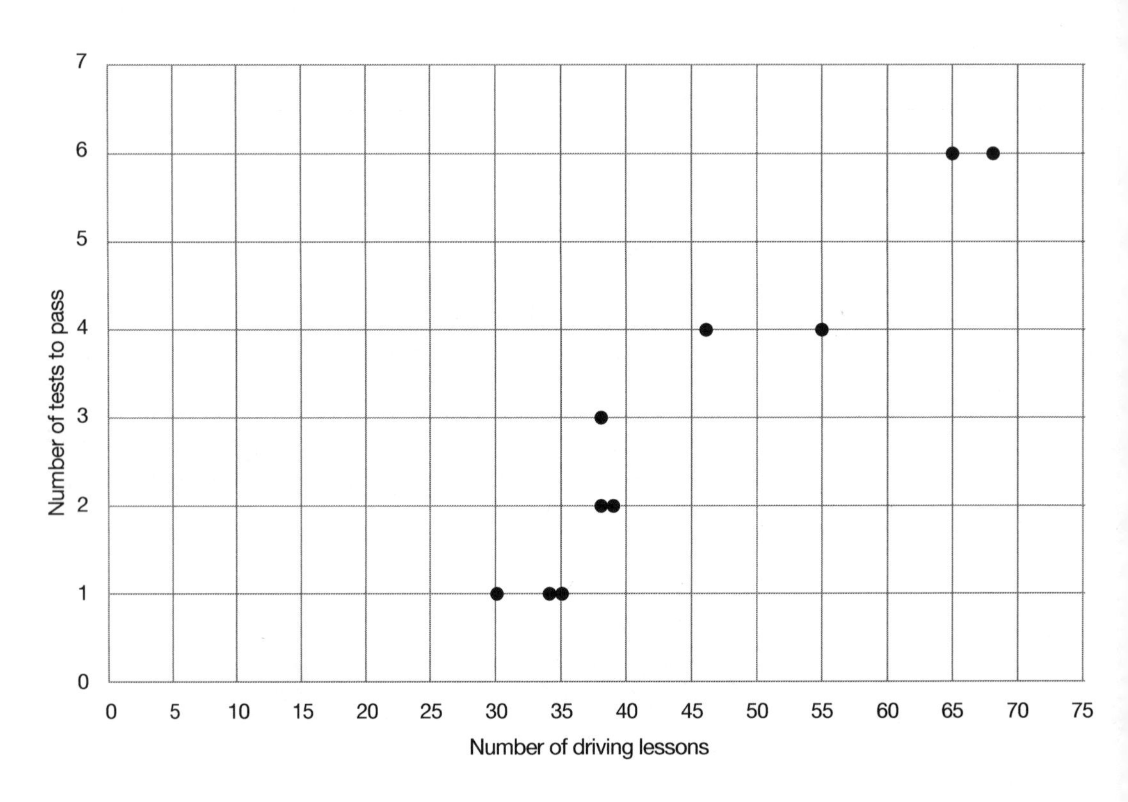

Answer

ANSWERS TO NUMERICAL REASONING – BASIC (Section 1)

Q1. 80

EXPLANATION = 120 (total number of people) ÷ 3 = 40. This is equal to 1/3. Therefore: 40 x 2 = 80.

Q2. 420

EXPLANATION = 700 ÷ 5 x 3 = 420.

Q3. 4

EXPLANATION = there are 1000 millilitres in 1 litre, therefore if there are 4000 millilitres, that will be equivalent to 4 litres.

Q4. 100°

EXPLANATION = the angle makes a straight line (which in essence, is a half turn of a circle). Therefore the angles would all need to add up to make 180°. So, 180 – 50 – 30 = 100°.

Q5. 1560

EXPLANATION = 120 x 13 = 1560.

Q6. £27

EXPLANATION = £45 ÷ 100 x 60 = £27.

Q7. 0

EXPLANATION = this shape is a parallelogram, and these shapes do not contain a line of symmetry. No matter where you draw the reflection line, the shape cannot be reflected symmetrically.

Q8. C = 1080 g

EXPLANATION = 120 x 9 = 1080 g. Pay attention to the measurements; the question is in grams (g), so therefore your answer should also be in grams, unless stated otherwise.

Q9. 324 cm²

EXPLANATION = the key thing to remember is that the shape is a square (the sides will be the same length). If the perimeter of the shape is 72cm, that means 72 needs to be divided by 4 (4 sides). So, $72 \div 4 = 18$. Each length of the side is 18 cm, and to work out the area = $18 \times 18 = 324$ cm^2.

Q10. ½

EXPLANATION = 24/48, both numbers can be divided by 24. It goes into 24 once, and goes into 48 twice. Therefore it gives the fraction of ½.

Q11. B = 19, 21

EXPLANATION = this is a series of repetition. The regular series adds 2 to every number.

Q12. B = 13, 21

EXPLANATION = this is a Fibonacci number sequence. The sequence follows the pattern of adding the two previous numbers together, to get the next number. For example, the 8 is found by adding the 5 and the 3 together.

Q13. 24:17

EXPLANATION = both amounts are in pounds. We have to convert both amounts into pence. £12.00 = 1200p. £8.50 = 850p. Now the ratio is 1200:850. Both sides are divisible by 50. Dividing both sides by 50 gives 24:17. So the ratio is 24:17.

Q14. 2:1

EXPLANATION = the answer is 2:1. You can divide both sides of 16:8 by 8.

Q15. 14

EXPLANATION: $6 \times 7 = 42 \div 3 = 14$.

Q16. 150

EXPLANATION = $120 \div 4 = 30 \times 5 = 150$.

Q17. 72

EXPLANATION = $88 \div 11 = 8 \times 9 = 72$.

Q18. C = 7 hours

EXPLANATION = total time spent marking one exam = 6 minutes (Literature) + 7 minutes (Language) + 2 minutes (checking) = 15 minutes. So, 28 exams will take = 15 (minutes) x 28 (exams) = 420 minutes. Converted into hours and minutes = 7 hours.

Q19. D = 90%

EXPLANATION = 0.9 x 100 = 90%.

Q20. B = 6x

EXPLANATION = x + 8x = 9x. So, 9x – 3x = 6x.

Q21. 169.6

EXPLANATION = 23.7 – 2.5 = 21.2 x 8 = 169.6.

Q22. 3 or 3/20

EXPLANATION = 20 – 12 – 5 = 3. So your chances of picking a white button is 3 out of a possible 20.

Q23. 18

EXPLANATION = 138 ÷ 8 = 17.25. You need one teacher for every 8 students, therefore you would need 18 members of staff in order to cater for 138 students.

Q24. The correct answer would have to look like this:

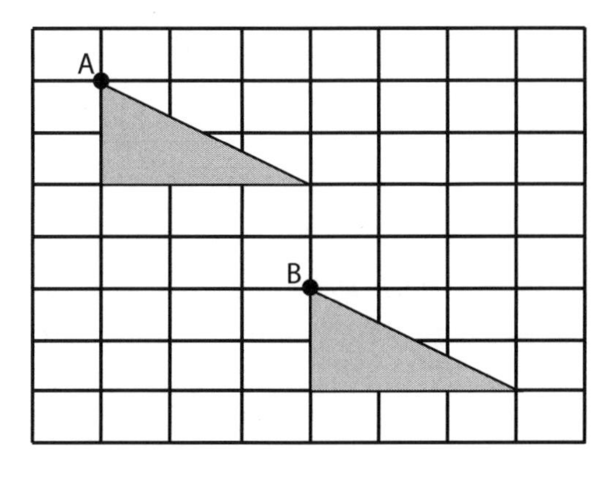

Q25. The correct answer would have to look like this:

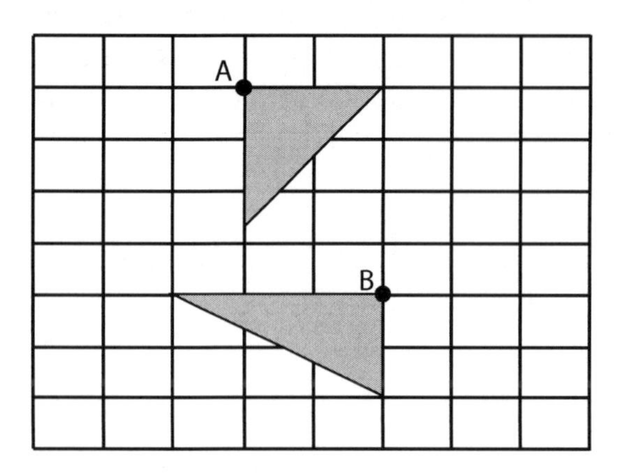

Q26. 09.25am

EXPLANATION = 0845 add 15 minutes (form time) = 9 o'clock. 9 o'clock add 25 minutes (assembly time) = 09.25am.

Q27. £6000

EXPLANATION = 27 rows of 28 seats = 756 – 5 (that are empty) = 751. 751 (number of seats) x £8 = £6008. To the nearest hundred = £6000.

Q28. D = 2500g

EXPLANATION = there are 1000g in 1 kilogram. Therefore, 2500g is equivalent to 2.5. (2.5 x 1000 = 2500).

Q29. A = 9/10

EXPLANATION = we use decimal points to distinguish whole parts from separate parts (tenths, hundredths, thousandths etc). A tenth is 1/10 of a unit, therefore the 9 represents 9 tenths of part of a unit.

Q30. 30% or 3/10 or 0.3

EXPLANATION = the set of data is for 10 people. 3 people passed first time so therefore: 10 ÷ 100 x 3 = 0.3 or 0.3 x 100 = 30% or simply 3 out of 10 (3/10).

Numerical Reasoning – ***BASIC***

(Section 2)

Question 1

Work out $\frac{2}{5} + \frac{7}{8}$

Answer

Question 2

Work out $\frac{4}{6} \times \frac{3}{5}$

Answer

Question 3

Below is a diagram of a cube. Work out its volume in cubic centimetres.

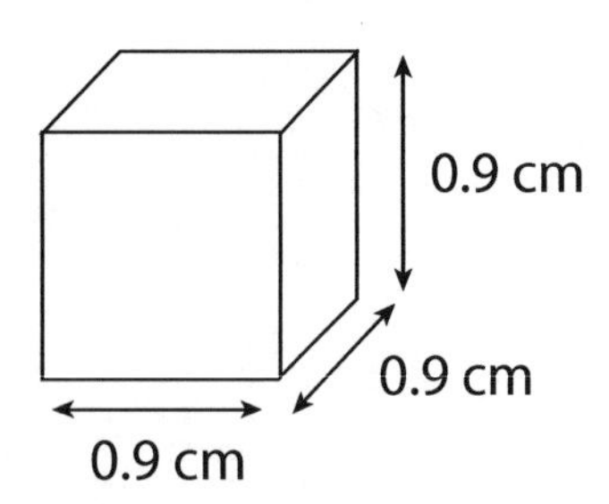

Answer

Question 4

Three whole numbers have a total of 100. The first number is a multiple of 15. The second number is ten times the third number. Work out the three numbers.

Answer

Question 5

The probability of picking a lottery winning ticket in the national lottery is 1 in 14 million. If 36 million tickets are sold weekly, how many jackpot winners, on average, would you expect in one week?

A	B	C	D
2 million	2	20	1

Question 6

A car travelled 100 metres in 9.63 seconds. On a second occasion, it travelled 200 metres in 19.32 seconds. Which distance had the greater average speed?

A	B	C	D
100 metres	200 metres	Both the same	Cannot say

Question 7

A function is represented by the following machine.

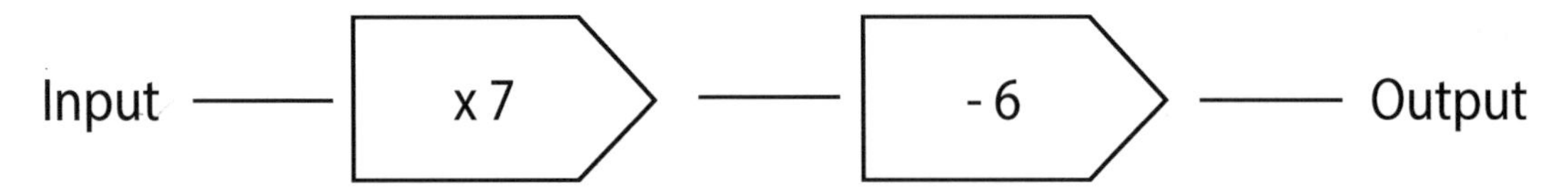

A number is put into the machine. The output of the machine is 71. What was the number first inputted into the machine?

Answer

Question 8

What is one quarter of 6 hours?

A	B	C	D
1 hour and 30 minutes	95 minutes	180 minutes	1 hour and 20 minutes

Question 9

Simplify 5w – 5x – 4w – 2x.

Answer

Question 10

A function is represented by the following machine.

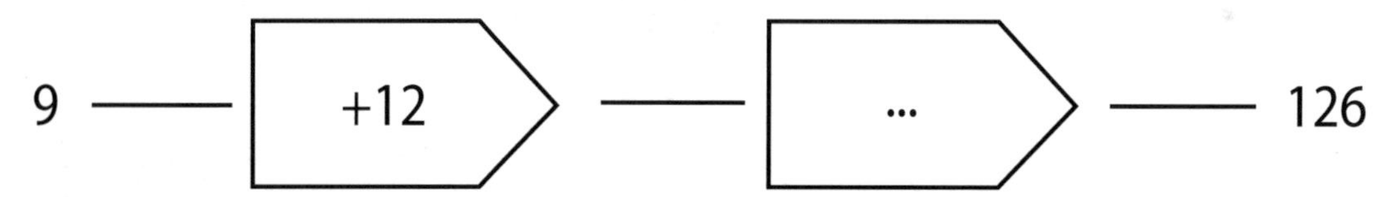

9 is put into the machine. The output of the machine is 126. What is the missing function in the second part of the machine sequence?

Answer

Question 11

Write down all the factors of 48?

Answer

Question 12

What month saw the mode number of pupils to be absent in the one month period, across all five subjects?

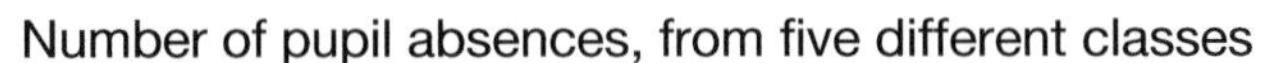

Number of pupil absences, from five different classes

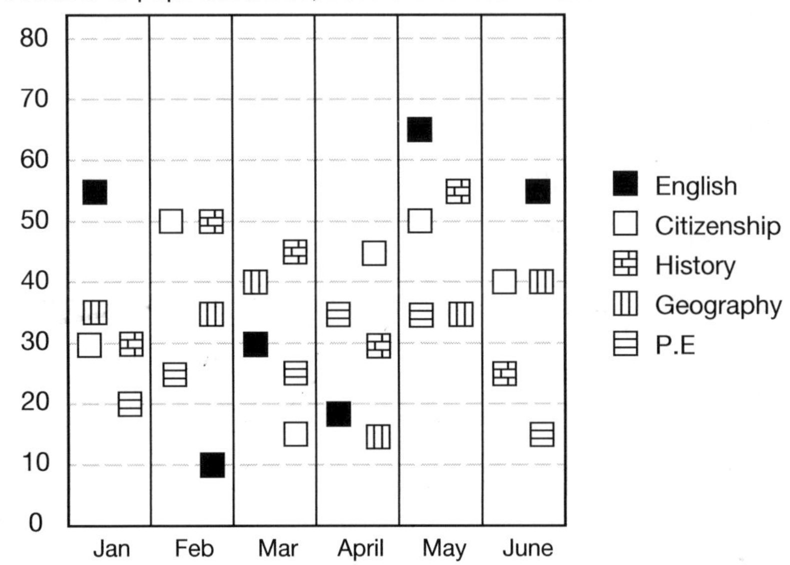

A	B	C	D
February	May	June	March

Question 13

How many different numbers can be made from these four playing cards?

4 9 3 1

Answer

Question 14

A Science exam is marked out of 50. There are 30 pupils in the class. The marks of the class are as follows:

7	**36**	**41**	**22**	**36**	**22**
41	**27**	**29**	**30**	**20**	**17**
9	**32**	**47**	**43**	**31**	**29**
27	**29**	**32**	**9**	**28**	**35**
17	**12**	**8**	**34**	**27**	**29**

Using this stem and leaf diagram, add the data in ascending order.

0	
1	
2	
3	
4	
5	

Question 15

Using the above stem and leaf diagram, what is the median?

Answer ______

Question 16

Work out the angles for A, B and C.

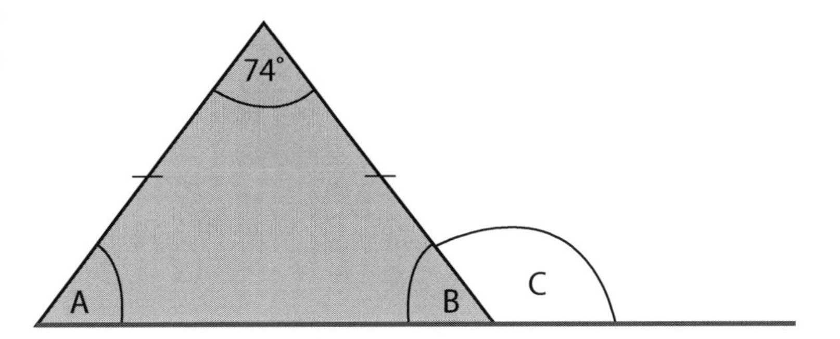

A =

B =

C =

Question 17

Calculate 158 x 67.

Answer

Question 18

A farmer has 630 eggs. They are to be placed in egg trays. Each tray can hold 36 eggs. How many trays will be needed to hold all of the eggs?

Answer

Question 19

Mark is going to make chocolate peanut squares. There are just three ingredients, chocolate, peanut butter and rice crispies, mixed in the ratio 4 : 2 : 3 respectively.

How much of each ingredient will he need to make 900 g of mixture?

Chocolate = [ ]

Peanut Butter = [ ]

Rice Crispies = [ ]

Question 20

Two of the numbers move from Box A to Box B. The total of the numbers in Box B is now four times the total of the numbers in Box A. Which two numbers move?

Box A	Box B
2, 6, 3, 9, 4	10, 1, 7, 8, 5

Answer [ ]

Question 21

Work out 256% of 6800.

Answer [ ]

Question 22

Subtract 3/8 of 104 from 5/7 of 98.

A	B	C	D
27	22	31	41

Question 23

Below is a pie chart illustrating the number of pupils studying a course in the following subject areas.

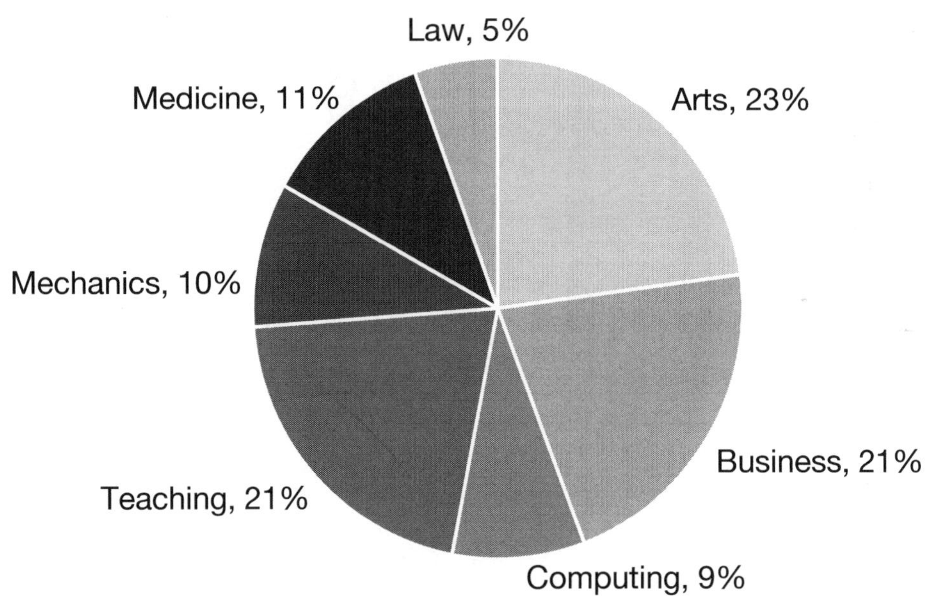

Students by Faculty

If the data is based on 3620 students, how many of those students are studying either mechanics or law?

Answer [ ]

Question 24

Below is a bar chart displaying some of the heights of the highest mountains.

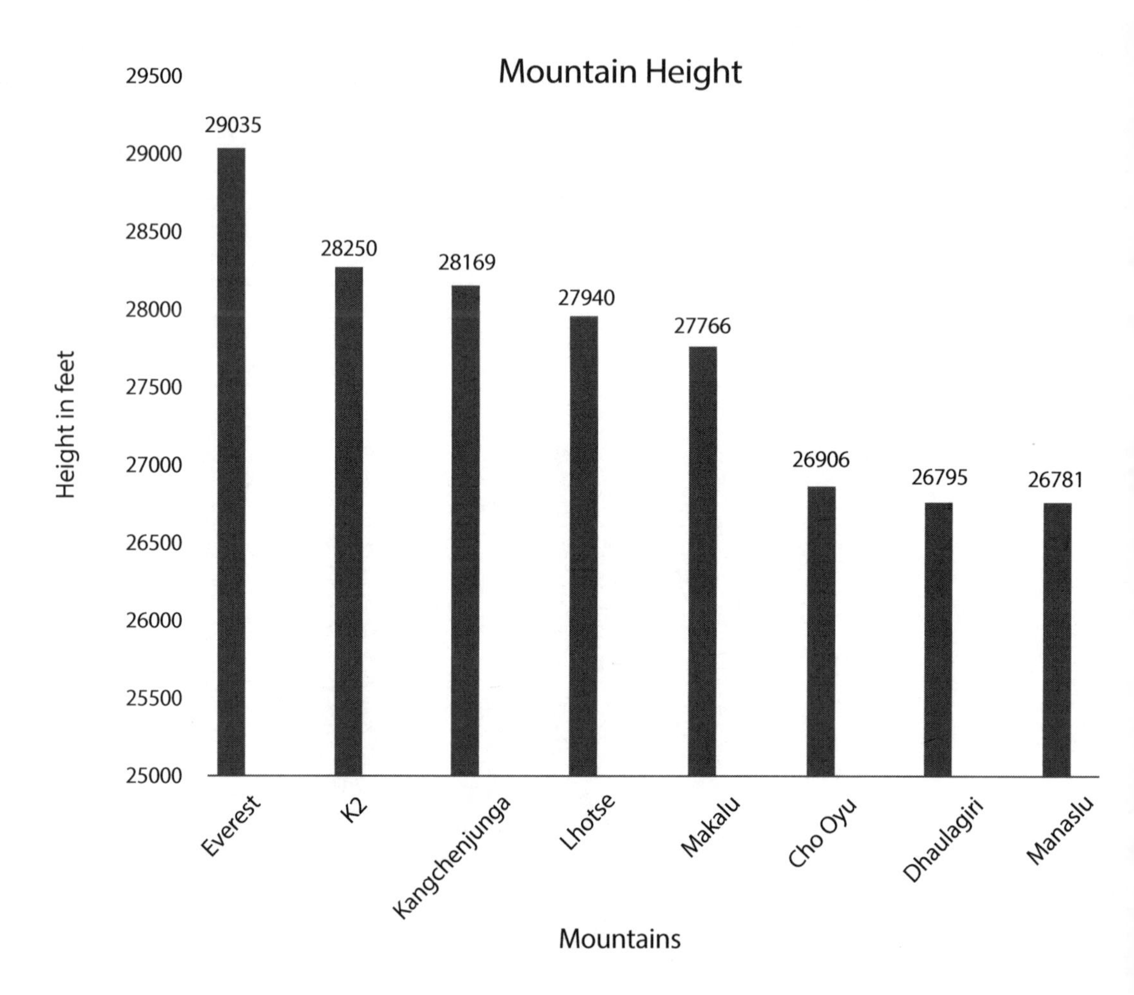

Work out the average height, to the nearest hundred, based on the information provided.

A	B	C	D
27,770	28,000	27,800	27,700

Question 25

What two numbers comes next in the sequence?

2, 4, 8, 16, 32, 64, … …

A	B	C	D
126 / 215	128 / 256	128 /265	182 /265

Question 26

Lisa cycles at an average speed of 8km/h. How far does she travel if she cycles for 4 hours?

Answer ☐

Question 27

James runs from 4.50pm until 5.20pm at an average speed of 7 km/h. How far did he go?

Answer ☐

Question 28

What is the highest common factor of 12 and 20?

A	B	C	D
4	8	12	2

Question 29

Here is a spinner. Circle the chance of the spinner landing on an odd number.

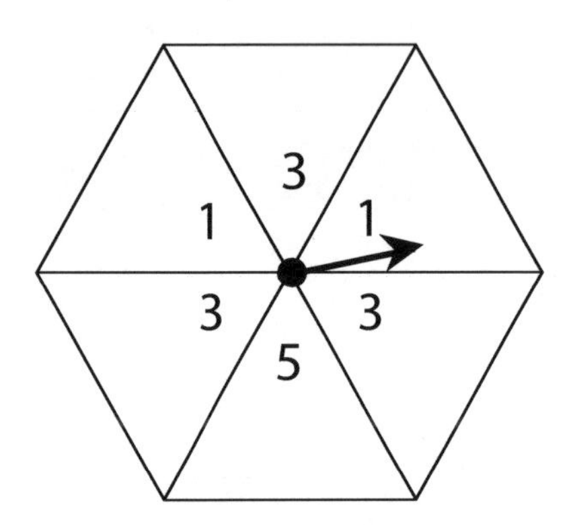

A	B	C	D
$\frac{6}{6}$ or 6	$\frac{4}{6}$	$\frac{1}{2}$	$\frac{1}{3}$

Question 30

What is the angle of D?

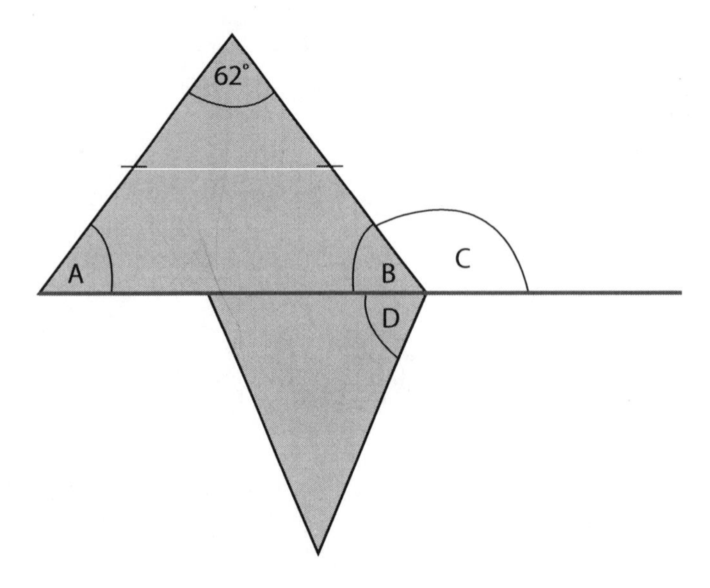

Answer

ANSWERS TO NUMERICAL REASONING – BASIC (Section 2)

Q1. $\frac{51}{40}$ or $1\frac{11}{40}$

EXPLANATION = $\frac{2}{5} + \frac{7}{8} = \frac{16 + 35}{40} = \frac{51}{40}$ or $1\frac{11}{40}$

Q2. $\frac{12}{30}$ $\frac{6}{15}$ or $\frac{2}{5}$

EXPLANATION = $\frac{4}{6} \times \frac{3}{5} = \frac{4 \times 3}{6 \times 5} = \frac{12}{30}$ or $\frac{6}{15}$ or $\frac{2}{5}$

Q3. 0.729

EXPLANATION = 0.9 x 0.9 x 0.9 = 0.729

Q4. 45, 50 and 5

EXPLANATION = three numbers with two of these criteria: a multiple of 15, two numbers in the ratio 10 : 1, and sum of 100.

- Multiples of 15 = 15, 30, 45, 60, 75, 90. So, the first number will be one of these numbers. Two of the numbers follow the rule of: being in the ratio 10 : 1. This works out to be 50 and 5, and will add up to 100 if you add the 45.

Q5. B = 2

EXPLANATION = 36 (million) ÷ 14 (million) = 2.57. So you could expect 2 lottery winners, on average, in a week.

Q6. A = 100 metres

EXPLANATION = 100 ÷ 9.63 = 10.384. 200 ÷ 19.32 = 10.351. Therefore 100 metres has the greatest average speed.

Q7. 11

EXPLANATION = 71 + 6 ÷ 7 = 11. (Remember, to work out the original number, you must work backwards. In order for you to work backwards, you must do the opposite to what the machine is telling you to do).

Q8. A = 90 minutes

EXPLANATION = 6 (hours) x 60 (minutes) = 360 minutes. So, 360 (minutes) ÷ 4 (1/4) = 90 minutes.

Q9. 1w – 7x

EXPLANATION = you need to break up the sequence: (5w) (-5x) (-4w) (-2x).

So, 5w – 4w = 1w and -5x – 2x = -7x. So this simplifies to: 1w – 7x.

Q10. Multiply by 6 or (x6)

EXPLANATION = 9 + 12 = 21. 126 ÷ 21 = 6. Therefore if you put (x6) into the equation (because you divided 126 by 6, you would put the opposite into the equation). Therefore, 9 + 12 x 6 = 126.

Q11. 1 and 48, 2 and 24, 3 and 16, 4 and 12, 6 and 8.

EXPLANATION = the definition of factors is 'all the numbers that can be divided into that number', i.e. what numbers can be multiplied together to reach that number?

Q12. B = May

EXPLANATION = you need to add up all of the subjects for each month. January = 170, February = 170, March = 155, April = 145, May = 240, June = 175. Therefore the mode (the most) in one given month is in May.

Q13. 24

EXPLANATION = starting with the number 4, you can get 6 numbers (4931, 4913, 4319, 4391, 4139, 4193). This can be done for all 4 numbers (if you start with a different number; you will be able to make 6 different numbers). Therefore 6 groups of 4 = 24.

Q14. Your answer should look exactly like this:

0	7 8 9 9
1	2 7 7
2	0 2 2 7 7 7 8 9 9 9 9
3	0 1 2 2 4 5 6 6
4	1 1 3 7
5	

Q15. 29

EXPLANATION = putting the data in order from smallest to biggest, you then need to find the median (middle) number. Do this by eliminating one number from the start, and one number from the end, until you reach the number in the middle. For this sequence, two numbers are left in the middle: 29 and 29. So, add both numbers and divide it by 2 to find the middle number. So, 29 + 29 ÷ 2 = 29.

Q16. A = 53°, B = 53°, C = 127°

EXPLANATION = a triangle contains 180°. So, 180 - 74° = 106°. Both A and B are going to be the same size (you will notice two small lines placed on both sides of the triangle, illustrating they're the same size and length). So, 106 ÷ 2 = 53°. To work out angle C, a straight line has 180°. You've just worked out angle B is 53°, so 180° – 53° = 127°.

Q17. 10,586

EXPLANATION = first, multiply by 7 (units): 158 x 7 = 1106. Then add a zero on the right side of the next row. This is because we want to multiply by 60 (6 tens), which is the same as multiplying by 10 and by 6. Now multiply by 6: 158 x 60 = 9480. Now add the two rows together: 9480 + 1106 = 10,586.

Q18. 18

EXPLANATION = 630 ÷ 36 = 17.5. So, you would need 18 trays in order to hold all the eggs.

Q19. Chocolate = 400g, Peanut Butter = 200g, Rice Crispies = 300g

EXPLANATION = to work out the chocolate: 900 ÷ 100 x 4 = 400g. To work out the peanut butter = 900 ÷ 100 x 2 = 200g. To work out the rice crispies = 900 ÷ 100 x 3 = 300g.

Q20. 9 and 4

EXPLANATION = if you moved 9 and 4, this leaves Box A with a total of 11. If you add 9 and 4 to 10, 1 7, 8 and 5, you will get 44. Therefore, this is 4 times as many.

Q21. 17,408

EXPLANATION = 6800 ÷ 100 x 256 = 17,408.

Q22. C = 31

EXPLANATION = 104 ÷ 8 x 3 = 39. 98 ÷ 7 x 5 = 70. So, 70 – 39 = 31.

Q23. 543

EXPLANATION = 3620 ÷ 100 x 5 = 181 (Law students). 3620 ÷ 100 x 10 = 362 (Mechanical students). So, the number of law and mechanical students is: 362 + 181 = 543.

Q24. D = 27,700

EXPLANATION = add up all of the sums and divide it by how many mountains there are (8). So, 29035 + 28250 + 28169 + 27940 + 27766 + 26906 + 26795 + 26781 = 221642 ÷ 8 = 27705.25. To the nearest hundred = 27,700.

Q25. B = 128 / 256

EXPLANATION = the sequence follows the pattern of 'the power of 2'. In other words, the number is multiplied by 2 each time. So, 64 x 2 = 128 and 128 x 2 = 256.

Q26. 32 km

EXPLANATION = Speed x time. So, 8 x 4 = 32 km.

Q27. 3.5 km

EXPLANATION = 4.50 pm – 5.20pm = 30 minutes. 30 minutes = 0.5 hour. Remember, distance = speed x time. So, distance = 7 x 0.5 = 3.5 km.

Q28. A = 4

EXPLANATION = the factors of 12 are: 1, 2, 3, 4, 6 and 12. The factors of 20 are: 1, 2, 4, 5, 10 and 20. So the highest common factor of 12 and 20 is 4.

Q29. A = 6/6 or 6.

EXPLANATION = the spinner contains only odd numbers. So no matter what number it lands on, you will always spin an odd number.

Q30. 59°

EXPLANATION = 180° - 62 = 118. 118 ÷ 2 = 59°. This would be the angle for angle A and B. It would also be the same for angle D, because B, D, C and the gap would need to make up 360°. So, 360 – 112 – 112 = 118. So both angles B and D are 59°.

Numerical Reasoning – ***INTERMEDIATE***

Our Numerical Reasoning (**intermediate**) section will provide you with the skills and knowledge expected for a strong GCSE level mathematical test. The difficulty of the questions will depend on the type of Numerical Reasoning test you take.

In order to help you pass a Numerical Reasoning test, we have provided you with lots of questions which will test you on an intermediate level.

In this type of intermediate Numerical Reasoning test, you can expect to find questions on the following areas:

- Percentages
- Fractions
- Decimals
- Data Interpretation
- Prime, multiple and factor numbers
- Mean / Mode / Median / Range
- Box and Whisker Plots
- Statistics
- Currency
- Mass / Density / Volume
- Stem and Leaf Diagrams

Whilst we have provided you an array of questions, your Numerical Reasoning test will be tailored to the job for which you are applying, and so the questions may not be the same, but will test the same skills and knowledge in terms of an intermediate level numerical test.

EXAMPLES of Numerical Reasoning - ***INTERMEDIATE***

Fractions / Decimals / Percentages

$$\frac{1}{10} = 0.1 = 10\%$$

How to work out fractions into decimals into percentages:

- 0.1 into a percent, you would move the decimal point two places to the right, so it becomes 10%.
- To convert 1/10 into a decimal, you would divide both numbers. For example, 1 ÷ 10 = 0.1.
- To convert 10% into a decimal, you move the decimal point two places to the left. For example, to covert 10% into a decimal, the decimal point moves two spaces to the left to become 0.1.

Number Sequences

5, 10, 15, 20, 25, 30, ... , ...

How to work it out:

In order to work out number sequences, you need to understand what is happening from one number to the next.

- For example, in the above number sequence: you will notice that each number is 5 higher than the previous.
- So, the next number after 30 should be 35.
- The next number after 35 should be 40.

Percent Increase

Work out the percentage increase.

Percent Increase

To work out the percentage increase of a set of data, you need to remember this formula:

Percent Increase % = Increase ÷ original number x 100

If your answer is a negative number, then this is a percentage **decrease**.

Percent Decrease

Work out the percentage decrease.

Percent Decrease

To work out the percentage decrease of a set of data, you need to remember this formula:

Percent Decrease % = Decrease ÷ original number x 100

If your answer is a negative number, then this is a percentage **increase.**

Stem and Leaf Diagrams

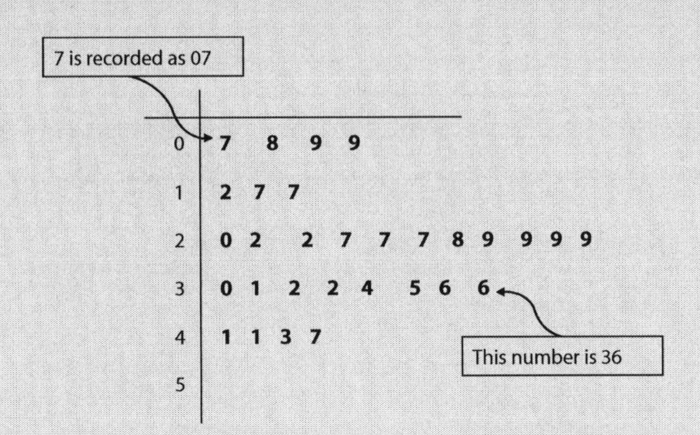

- Stem and leaf diagrams act as a way of handling data.
- These become particularly useful when dealing with large sums of data.
- They are also helpful ways to work out the mean, **mode**, **median** and **range**.

Mean / Mode / Median / Range

Mean

- To work out the mean of a set of data, you add up all the numbers and divide it by how many there are.

Mode

- The mode is easily remembered by referring to it as the 'most'. What number occurs most throughout the data?

Median

- Once the data is in ascending order, you can then work out what number is the median; what number is in the middle? If no number is in the middle, use the two numbers: add them up and divide by 2.

Range

- In ascending order, the range is from the smallest number to the biggest number.

Mass / Density / Volume

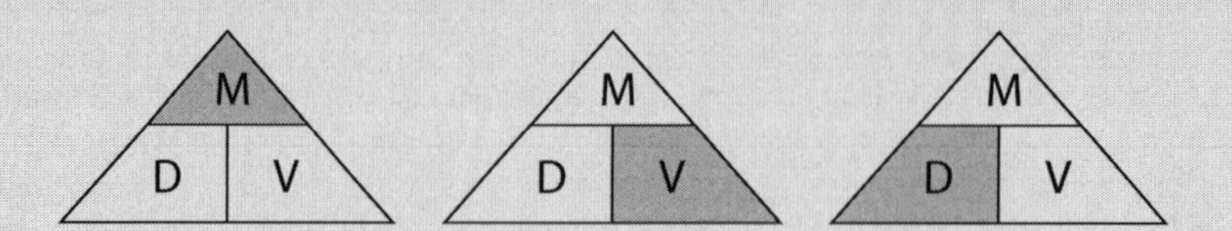

To work out the Mass:

- Mass = Density x Volume

To work out the Volume:

- Volume = Mass ÷ Density

To work out the Density:

- Density = Mass ÷ Volume

Ratio

There are 20 girls and 12 boys in a class. What is the ratio of girls to boys? Give your answer in its simplest form.

How to work it out:

- To work out the ratio in its simplest form, you need to find a number that both numbers can be divided by, until it can no longer be divided.
- So, for the above example:
    - o Start with 2, can both numbers be divided by 2? Yes, this will give you the new ratio of 10 : 6.
    - o Can both numbers be divided by 2 again? Yes, this will give you the new ratio of 5 : 3.
    - o This ratio is now in its simplest form, because the numbers 3 and 5 are prime numbers (can only be divided by the number itself and one).

Box and Whisker Plots

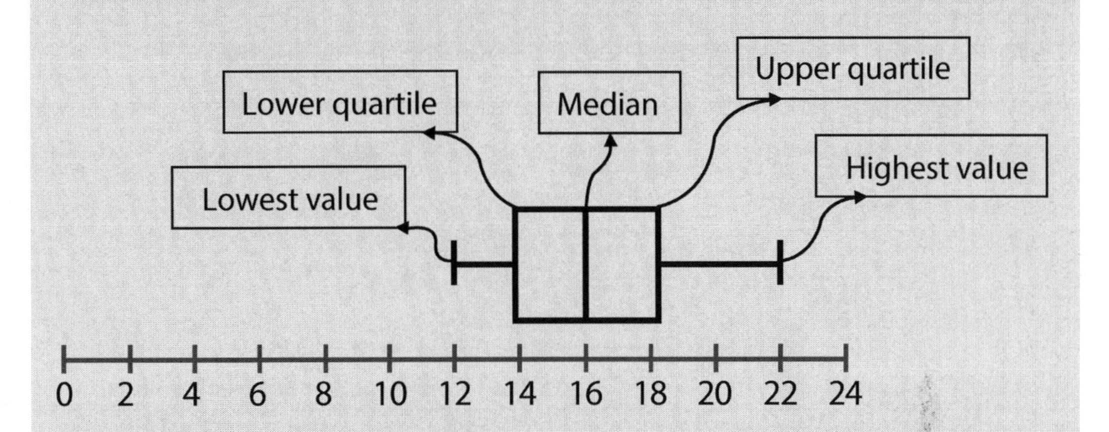

- From the above box and whisker plot, the value of each are as follows:
  - o Lowest value = 12
  - o Lowest quartile = 14
  - o Median = 16
  - o Upper quartile = 18
  - o Highest value = 22
- The use of box and whisker plots are to help evaluate a set of data and determine the range and quartiles of information.

Hectares

Work out the area of the shape. Write your answer in hectares.

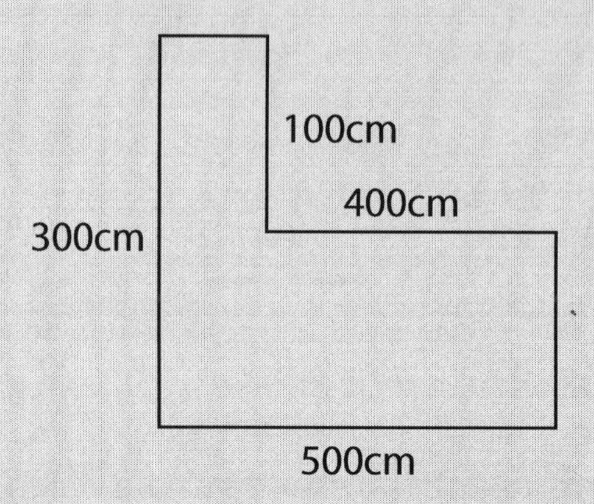

Hectares

The important thing to remember when dealing with hectares, is to use this information as a guideline:

1 hectare = 10,000 m = 2.47 acres

So, if you had measurements in centimetres, you would need to convert them into metres before attempting to convert the measurements into hectares.

Example

For the above example, to work out the area of the shape in hectares, you need to work out the area of the shape first, and then convert the centimetres into metres.

Step 1 = 300 x 500 = 150,000 cm = 150,000 ÷ 100 = 1500m

Step 2 = 100 x 400 = 40,000 cm = 40,000 ÷ 100 = 400m

Step 3 = 1500 – 400 = 1100

Step 4 = to convert 1100m^2 into hectares, we know 10,000 m^2 = 1 hectare. So, 1100 ÷ 10,000 = 0.11 hectares.

Inputs and Outputs

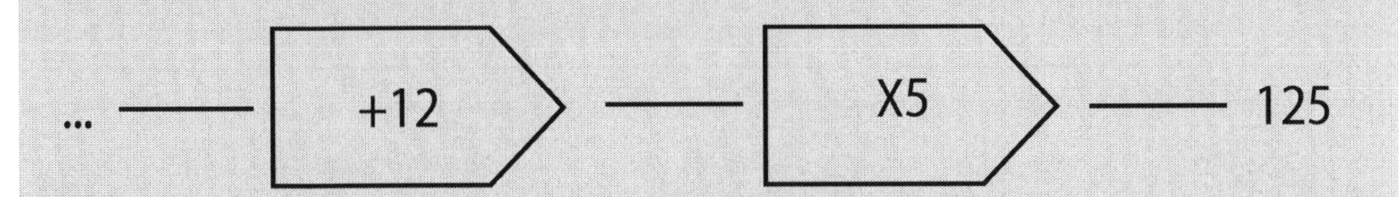

How to work it out:

In order to work out the missing number at the start of the sequence, you will need to work backwards.

- When working backwards, you need to do the OPPOSITE.
- For example:
  - o $125 \div 5 - 12 = 13$

Simplifying Equations

Simplify $4w - 6x - 2w - 1x$

$$(4w)\ (-6x)\ (-2w)\ (-1x)$$

$$(4w - 2w) = 2w$$

$$(-6x - 1x) = -7x$$

$$2w - 7x$$

- The important thing to remember for simplifying equations is to break up the equation (like above).
- The '-' signs and the '+' signs should also be grouped and be on the left of the number.

Numerical Reasoning – ***INTERMEDIATE***

(Section 1)

Study the following chart and answer the four questions that follow.

Bike sales

Country	Jan	Feb	Mar	April	May	June	Total
UK	21	28	15	35	31	20	150
Germany.	45	48	52	36	41	40	262
France	32	36	33	28	20	31	180
Brazil	42	41	37	32	35	28	215
Spain	22	26	17	30	24	22	141
Italy	33	35	38	28	29	38	201
Total	195	214	192	189	180	179	1149

Question 1

What percentage of the overall total was sold in April?

A	B	C	D	E
17.8%	17.2%	18.9%	16.4%	21.6%

Question 2

What percentage of the overall total sales were bikes sold to the French importer?

A	B	C	D	E
15.7%	18.2%	18.9%	25.6%	24.5%

Question 3

What is the average number of units per month imported to Brazil over the first 4 months of the year?

A	B	C	D	E
28	24	32	38	40

Question 4

What month saw the biggest increase in total sales from the previous month?

A	B	C	D	E
January	February	March	April	May

Study the following chart and answer the four questions that follow.

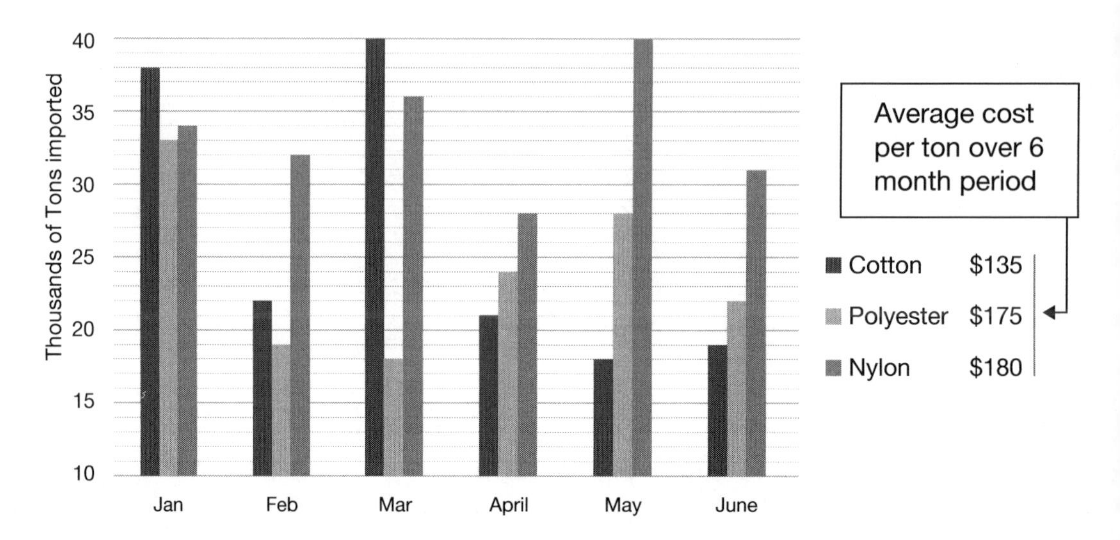

Question 5

What is the mean value for nylon imported over the 6 month period?

A	B	C	D	E
42.5	18.5	33.5	49.5	37.5

Question 6

What is the range for polyester imports across the 6 month period?

A	B	C	D	E
15	21	23	52	51

Question 7

What was the difference in thousands of tons between, cotton material and nylon material imports in the first 3 months of the year?

A	B	C	D	E
5	15	24	17	2

Question 8

What was the approximate ratio of polyester and nylon material imports in the first 4 months of the year?

A	B	C	D	E
94:120	94:130	92:110	95:100	94:90

Question 9

There are 60 girls and 65 boys in the lunch hall at school. What is the ratio of girls to boys? Give your answer in its simplest form.

Question 10

Look carefully for the pattern, and then choose which pair of numbers comes next.

1, 3, 6, 10, 15, 21, 28

A	B	C	D	E
42, 56	42, 48	30, 36	32, 36	36, 45

Question 11

The lowest percentage for attendance in Year 7 was 51%. The highest attendance was 100%. The median percent for attendance is 70%. The lower quartile percent was 61% and the upper quartile percent was 90%. Represent this information with a box-and-whisker plot.

Question 12

What is the amount of the lower quartile?

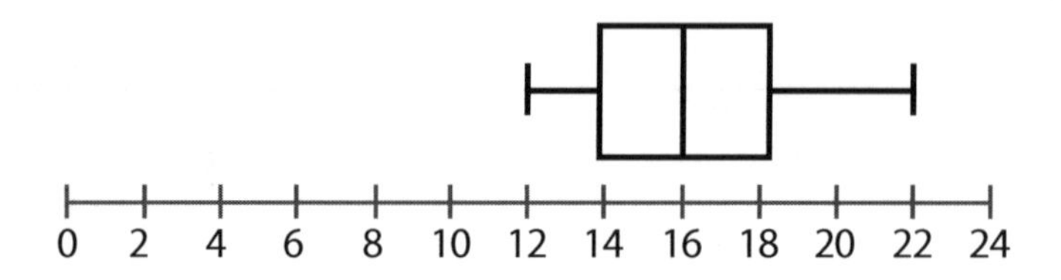

Answer

Question 13

The set of data below shows the results in a year 11 Media mock exam. The marks are out of 100%. The teacher wants to find the mean mark for this test which was given to 68 pupils. Give your answer to 1 decimal place.

Media mock exam (%)	No. of pupils	No. of pupils X media mock exam (%)
10	0	10 x 0 = 0
20	2	20 x 2 = 40
30	3	
40	6	
50	8	
60	11	
70	8	
80	15	
90	12	
100	3	
Totals	68	

The mean mark is:

Question 14

The two way table compares pupils' results for GCSE English with GCSE Media grades.

	Media GCSE Grades								
English GCSE Grades	A*	A	B	C	D	E	F	U	Total
A*									
A		2	2	3					7
B		1	3	4				1	9
C			8	10	6	1			25
D				1		2			3
E								1	1
F									
U									
Total		3	13	18	6	3		2	45

The percentage of pupils who received a D grade in Media is approximately what? To the nearest whole number.

Answer [ ]

Question 15

Below is a stem and leaf diagram showing the finishing time, in seconds, of 15 sprinters who took part in a race.

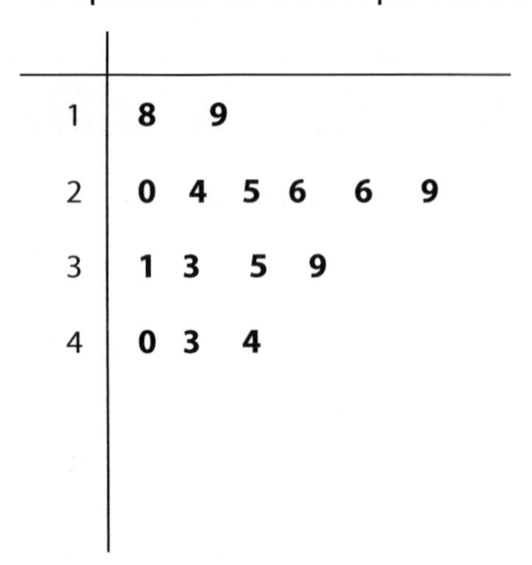

1 | 8 9
2 | 0 4 5 6 6 9
3 | 1 3 5 9
4 | 0 3 4

What is the median finishing time?

Answer

Question 16

Using the above stem and leaf diagram, what is the mean finishing time? To one decimal point.

Answer

Question 17

A ruler is 30 cm in length, correct to the nearest centimetre. What is the smallest possible length of the ruler?

Answer

Question 18

The head of English created the following table showing the number of pupils in each year group who got a C grade or above in their test.

Year Group	No. of pupils	No. of pupils who achieved a C grade or above in their English test
7	86	56
8	93	48
9	102	72
10	99	52
11	106	85
12	68	56

What is the percentage of pupils in all the year groups combined that got a C grade or above in their test. Give your answer rounded to a whole number.

Answer

Question 19

Add 7/9 of 189 to 5/8 of 128.

Answer

Question 20

Mark believes he knows what his brother Ryan is thinking. He carries out an experiment to test this. Mark and Ryan sit back-to-back. Ryan rolls an ordinary fair dice. Ryan then thinks about the number on the dice while Mark tries to predict this number.

In 400 attempts, how many correct predictions would you expect Mark to make if he was just guessing? Write your answer as a percentage of the 400 attempts.

Answer [ ]

Study the following table and answer the following questions.

Company	Company Profit (Annual) (£)	Cost to buy company (£)	Number of employees
A	15,000	18,000	6
B	26,000	24,000	11
C	22,000	20,000	8
D	40,000	40,000	10

Question 21

Using the above table, which company has the lowest annual profit per employee?

A	B	C	D	E
Company A	Company B	Company C	Company D	Company C and D

Question 22

Using the above table, approximately how many more employees would company C have to employ to achieve annual profit of £44,000?

A	B	C	D	E
4	11	8	3	19

Question 23

Using the above table, if company A makes an annual profit of £31,000 the following year, what is the percentage increase?

A	B	C	D	E
106.6%	94.5%	6.6%	51.6%	103.2%

Question 24

Using the above table, if company D makes an annual profit of £15,000 the following year, what is the percentage decrease?

A	B	C	D	E
105.6%	62.5%	33.5%	101.25%	71%

Study the following line graph and answer the following questions.

Ratio of exports and imports for two manufacturing companies

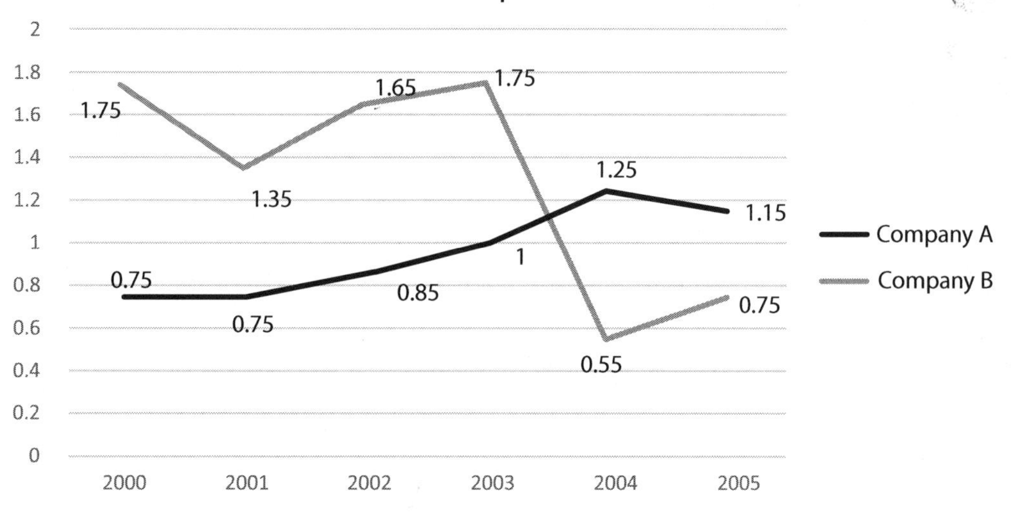

Question 25

Using the graph above, what is the difference between the number of imports and exports for company A in 2000, compared to the number of imports and exports for company B in 2005?

A	B	C	D	E
0	0.1	0.25	0.75	3

Question 26

Using the graph above, in how many of the years were the imports for Company B, more than their exports?

A	B	C	D	E
1	2	3	4	5

Question 27

Using the graph above, in what year was the difference between the imports to exports the maximum for Company B?

A	B	C	D	E
2000	2001	2002	2003	Cannot be determined

Question 28

Using the graph above, if the exports for Company A in 2002 were 250, what was the amount of imports in that year?

A	B	C	D	E
220	196	294	274	312

Below is a table representing the actual and target income for 2001 for five different companies.

COMPANY	ACTUAL INCOME (ANNUAL) for 2001	TARGET INCOME (ANNUAL) for 2001
Company A	£234,570	£300,000
Company B	£420,000	£421,560
Company C	£215,750	£450,000
Company D	£310,250	£325,000
Company E	£375,995	£325,000

Question 29

Using the above table, in the following year, Company B earns £275,000. What is the percentage decrease from Company B's earnings in 2001 and the earnings in the following year?

A	B	C	D	E
65.5%	45.5%	39.5%	34.5%	41.5%

Below is a table representing costs for different products.

PRODUCT	No. of units (1000's)	Cost of material per unit	Manu-facturing costs per unit	Total cost per unit	Sales price per unit	Total Sales Revenue After tax
P	18.5	2.25	2.15	4.4	6.5	38,850
Q	29.5	4.75	2.25	7	10	?
R	9	1.5	1.75	3.25	8	42,750
S	13	3.5	2.25	5.75	15	120,250

Question 30

Using the above table, product P gets taxed 25% of the total sales revenue. What would the total sales revenue be before tax?

A	B	C	D	E
£43,126.75	£48,215.25	£48,562.50	£49,100.05	£47,251.25

ANSWERS TO NUMERICAL REASONING – INTERMEDIATE (Section 1)

Q1. D = 16.4

EXPLANATION = to work out the percentage overall total that was sold in April, divide how many bikes were sold in April (189) by the total (1149) and then multiply it by 100. (189 ÷ 1149 x 100 = 16.4).

Q2. A = 15.7%

EXPLANTATION = to work out the overall percentage total that was sold to France, divide how many bikes were sold to France (180) by the total (1149) and then multiply it by 100. (180 ÷ 1149 x 100 = 15.66). Rounded up to 1 decimal place = 15.7.

Q3. D = 38

EXPLANTATION = to work out the average number of units per month imported to Brazil over the first 4 months of the year, you add up the first 4 amounts (Jan-April) and then divide it by how many numbers there are (4). So, (42 + 41 + 37 + 32 = 152 ÷ 4 = 38).

Q4. B = February

EXPLANATION = to work out the biggest increase in total sales from the previous month, you work out the difference between the totals for each of the month and work out which has the biggest increase. Between January and February, there was an increase by 19. None of the other months have a bigger increase and therefore February is the correct answer.

Q5. C = 33.5

EXPLANATION = nylon material = 34 + 32 + 36 + 28 + 40 + 31 = 201 ÷ 6 = 33.5.

Q6. A = 15

EXPLANATION = to work out the range, find the smallest and highest number of polyester imports (18) and (33) So, 33 – 18 = 15 (thousands).

Q7. E = 2

EXPLANATION = to work out the difference, add up the first 3 months for cotton (38 + 22 + 40 = 100). Add up the first 3 months for nylon (34 + 32 + 36 = 102). So, the difference between cotton and nylon = 102 – 100 = 2 (thousands).

Q8. B = 94:130

EXPLANATION = 94,000:130,000. Divide both numbers by 1000 to give you 94:130.

Q9. 12 : 13

EXPLANATION = the ratio of girls to boys is 60:65. However, both sides of this ratio are divisible by 5. Dividing by 5 gives 12:13. 13 has no common factors (apart from 1). So the simplest form of the ratio is 12:13. This means there are 12 girls in the lunch hall for every 13 boys.

Q10. E = 36 and 45

EXPLANATION = the sequence follows the pattern of adding by an extra number each time. For example +2, +3, +4, +5. Therefore, we would need to add 8, and then 9 to get the next two numbers. 28+8=36, and 36+9=45.

Q11. Your box and whisper plot diagram should look like this:

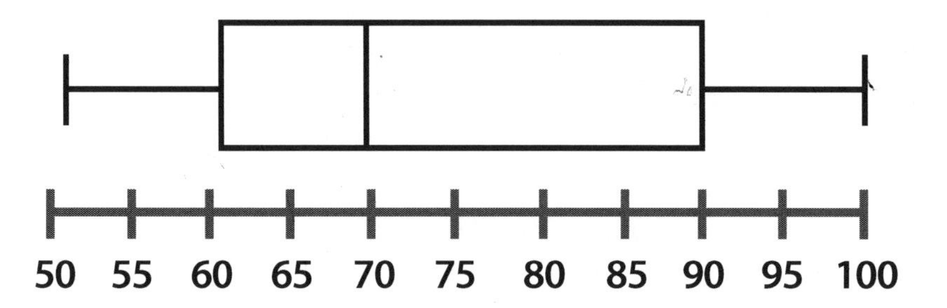

Q12. 14

EXPLANATION = The lowest quartile range is the first line that forms the box.

Q13. 67.2%

EXPLANATION = add up the "number of pupils multiplied by media mock exam" and then divide it by the "number of pupils".

Media mock exam (%)	No. of pupils	No. of pupils X media mock exam (%)
10	0	10 x 0 = 0
20	2	20 x 2 = 40
30	3	30 x 3 =90
40	6	40 x 6 = 240
50	8	50 x 8 = 400
60	11	60 x 11 = 660
70	8	70 x 8 = 560
80	15	80 x 15 = 1200
90	12	90 x 12 = 1080
100	3	100 x 3 = 300
Totals		

So, 4570 ÷ 68 = 67.2%.

Q14. 13%

EXPLANATION = number of pupils who received a D grade in Media = 6.

Total number of pupils = 45.

So, 6 ÷ 45 x 100 = 13.333%. To the nearest whole number = 13%.

Q15. 29 seconds

EXPLANATION = 'median' simply means 'middle'. So, what number is in the middle? Using the data in ascending order, you will notice that 29 (seconds) is the median/middle number.

Q16. 30.1 seconds

EXPLANATION = to work out the mean number, add up all the numbers and then divide it by how many numbers there are.

So, 452 ÷ 15 = 30.133. To one decimal point = 30.1.

Q17. 29.5 cm

EXPLANATION = if 29.5 is rounded up to the nearest whole number, it becomes 30cm. If the number is less than 29.5, like 29.4, it would be rounded down to 29cm. Therefore, 29.5cm is the smallest possible length the ruler can be.

Q18. 67%

EXPLANATION = add up total number of pupils = 554.

Add up the number of pupils who achieved a C grade or above in English = 369.

To work out the overall percentage = 369 ÷ 554 x 100 = 66.6%.

To the nearest whole number = 67%.

Q19. 227

EXPLANATION = 189 ÷ 9 x 7 = 147.

128 ÷ 8 x 5 = 80.

So, 80 + 147 = 227.

Q20. 16.75%

EXPLANATION = In to order to work out an estimate of how many times Mark will predict the correct number, you need to divide the number of attempts, by the number of probability (6). You know the probability is 6 because there is a 1/6 chance of rolling the same number.

So, 400 ÷ 6 = 66.666%. To the nearest whole number = 67%. You then need to work out 67% as a percentage of 400. So, 67 ÷ 400 x 100 = 16.75%.

Q21. B = Company B

EXPLANATION = simply divide the annual profit for each company by the number of employees, and see which company has the lowest profits.

Q22. C = 8

EXPLANATION = 44,000 ÷ 2750 = 16. That is 8 more than what they have already.

Q23. A = 106.6%

EXPLANATION = 31,000 – 15,000 = 16,000.

So, 16,000 ÷ 15,000 x 100 = 106.6%.

Q24. B = 62.5%

EXPLANATION = 40,000 – 15,000 = 25,000.

So, 25,000 ÷ 40,000 x 100 = 62.5%.

Q25. A = 0

EXPLANATION = the number of imports and exports for company A in 2000 = 0.75. The number of imports and exports for Company B in 2005 = 0.75. Difference = 0.

Q26. D = 4

EXPLANATION = if the ratio export to import is greater than 1, it means that the export was more than the import. For company B, the years 2000, 2001, 2002 and 2003 all have more than 1.

Q27. E = cannot be determined

EXPLANATION = it cannot be determined based on the information provided. You do not have any statistical data to work out the difference between exports and imports.

Q28. C = 294

EXPLANATION = 250 (the amount you are working out) ÷ 0.85 (exports in 2002 for Company A) = 294.

Q29. D = 34.5%

EXPLANATION = 420,000 – 275,000 = 145,000.

So, 145,000 ÷ 420,000 x 100 = 34.5%.

Q30. C = £48,562.50

EXPLANATION = 38,850 ÷ 100 x 25 = 9712.5.

So 38,850 + 9712.50 = 48,562.50.

Numerical Reasoning – ***INTERMEDIATE***

(Section 2)

The number of times a die was cast and the number of times each individual number appeared

Casts	1	2	3	4	5	6
First 10	2	3	1	1	2	1
First 20	5	4	3	4	3	1
First 30	8	5	6	5	4	2
First 40	10	6	7	6	5	6
First 50	13	7	10	7	6	7

Question 1

The same number did not appear on any two consecutive casts. If the number 4 appeared in the 20th cast, which number/s could not have appeared in the 11th cast?

A	B	C	D	E
4	1	2 and 3	6	3

Question 2

Using the above table, which of the following numbers must have turned up the least amount of times in the first 50 casts?

A	B	C	D	E
2, 4 and 6	6	5	3	1

Question 3

Using the above table, if the same number occurred for the 43rd cast and the 47th cast, what number/s could it be?

A	B	C	D	E
1	2	4	1 and 6	6

Question 4

Using the above table, what is the total ratio of the number of 3's cast and the number of 5's cast? In its simplest terms.

A	B	C	D	E
10 : 13	5 : 3	7 : 10	10 : 6	3 : 5

Representation of the grades students achieved across five subjects

	English	Maths	Science	History	Media
David	A-	B+	C-	C+	B+
Billy	C-	C+	B+	A+	A
Elliott	B+	B-	A+	A-	C
Taralyn	C+	B+	B+	C+	A+
Alecia	C	C+	A-	B-	C+
James	B-	B+	C-	C+	C
Gareth	B+	B-	A	B-	C-
Duncan	B-	C-	C+	C-	C
Joe	B+	B	B	C	A

Grade	Pass Mark
A+	96-100
A	91-95
A-	86-90
B+	81-85
B	76-80
B-	71-75
C+	65-70
C	59-64
C-	50-58

Question 5

In the above table, find the minimum possible of total marks for all nine can- in Science?

A	B	C	D	E
)7	776	667	676	None of these

ion 6

bove table, what is the highest mark across all five subjects that Da- d have got?

	B	C	D	E
8	386	320	408	None of these

Question 7

In the above table, what would Gareth's total average mark be, if he had scored the average mark in all his subjects?

A	B	C	D	E
78	75.2	76.2	79.4	76.5

Question 8

In the above table, if everyone scored the lowest pass mark in each subject, who scored the best overall?

A	B	C	D	E
David	Taralyn	Alecia	Elliott	Joe

Question 9

There are two lists of numbers. One list contains 11 numbers, the average of which is 36. The second list contains 13 numbers and has the average of 41. If the two lists are combined, what is the average of the numbers in the new list? To the nearest whole number.

A	B	C	D	E
36	37	38	39	40

Question 10

A farmland area is measured to be 220m in length by 80m in width. What is the approximate area of the field in hectares? 1 hectare = 10,000m2 = 2.47 acres.

A	B	C	D	E
17 hectares	1.76 hectares	176 hectares	17.6 hectares	7.67 hectares

Question 11

The sterling to US dollar rate is 1:1.32. How many dollars would you receive if you changed up £450?

A	B	C	D	E
$592	$541	$594	$531	$441

Question 12

A flight leaves the airport at 2200 hours. It is an 11 hour and 45 minute flight. There is a 2 hour time difference. What is the time they arrive at their destination, assuming the time difference is 2 hours in front?

Answer

Question 13

Multiply 0.04 by 1.1.

Answer

Question 14

What is $6/8 \div 2/3$? Write your number using mixed fractions and its simplest form.

Answer

Question 15

David walks at a pace of 0.5 metres per second. Assuming his pace does not change and he walks continuously during this time, how far will David walk after 8 minutes? Give your answer in centimetres.

Answer

Question 16

James is going on holiday and has a suitcase to check in. His luggage cannot exceed 15 kg or he will have to pay an additional charge. If 45% of the total allowance is accounted for by clothing, 21% of the total allowance is accounted for by shoes and 7% of the total allowance is accounted for by toiletries; how much of the 15 kg allowance does he have to spare? Give your answer in kg.

Answer

Question 17

A lady has been prescribed medication by her doctor. She is prescribed a 10.5 fluid ounce bottle of medication with the instructions to take 0.25 fluid ounces three times a day. How many days does she have to take the medication for?

A	B	C	D	E
7 days	10 days	12 days	13 days	14 days

Question 18

What is the volume of this object?

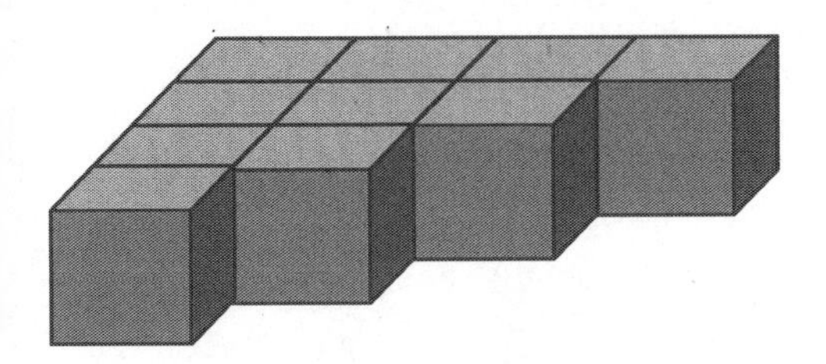

A	B	C	D
20 cubic units	10 cubic units	11 cubic units	15 cubic units

Question 19

What is 2/8 x 9? Simplify your answer and write it as a mixed fraction.

A	B	C	D
2 1/3	1 1/4	2 2/8	2 1/4

Question 20

What is the prime factorisation of 9?

A	B	C	D
3	3 x 9	3 x 3	3 x 3 x 3

Question 21

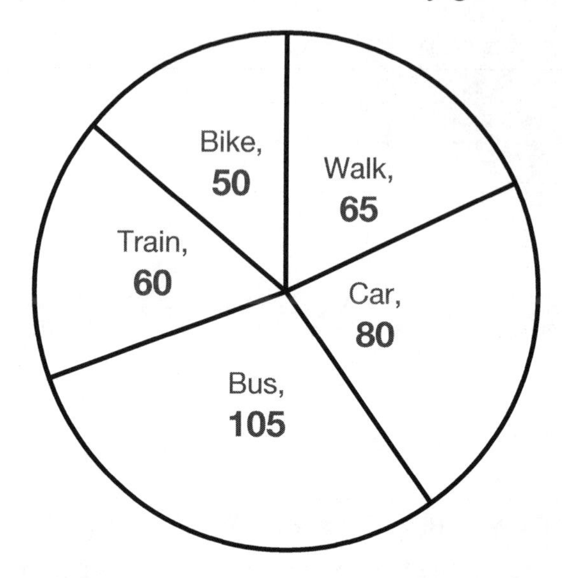

Among the respondents, 85% of the people who said they walk to school and 90% of the people who said they bike to school also said that their school was less than 5 miles away. How many people said that they walk or bike to school because it is less than 5 miles away? Rounded to the nearest whole number.

A	B	C	D	E
50	100	120	125	145

Question 22

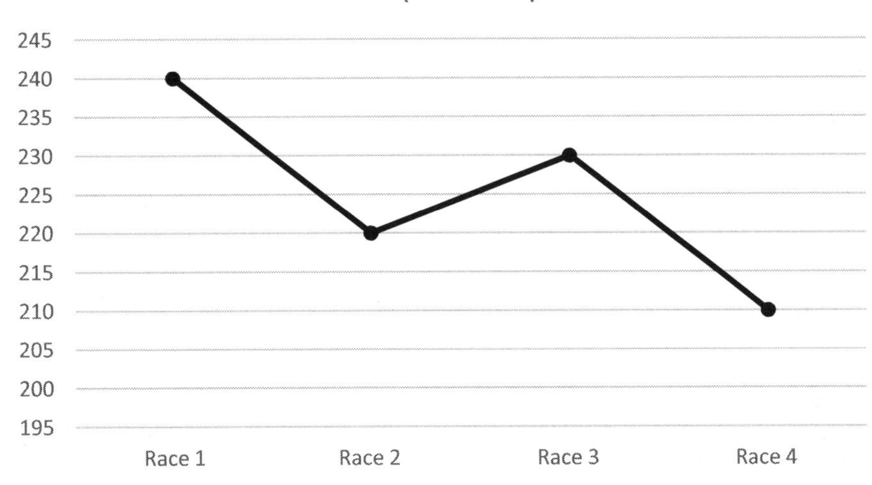

Harrison trains before his next race. In race 5, Harrison is able to run the 800m in 180 seconds. What is the difference between his time for Race 5, and his overall average time before he began training?

A	B	C	D	E
45	35	38	32	62

Question 23

Work out the density of a 9 kg lump of metal with a volume of $2.65m^3$. Write your answer to two decimal places.

Answer

Question 24

You have a rock with a volume of 30cm3 and a mass of 60g. What is its density?

Answer

Question 25

You decide you want to carry a boulder home from the beach. It is 30 centimetres on each side, and has a volume of 27,000cm³. It is made of granite, which has a typical density of 2.8 g/cm³. How much will this boulder weigh?

Answer

Question 26

A History test consists of three papers. The first paper will be marked out of 80, and has a weighting of 35% towards the final grade. The second paper will be marked out of 60, and has a weighting of 30% towards the final grade. The third and final test will be marked out of 40, and has a weighting of 35% towards the final grade. A pupil scores 60 in the first paper, 45 in the second, and 35 in the third paper. What is the pupil's final percentage score? Give your answer to 1 decimal place.

Answer

Question 27

There are 45 students in a class. 37 students have completed their challenge of reading 6 books. What is the percentage number of students who have completed their reading challenge? Round up to the nearest whole number, and give your answer to 1 decimal place.

Answer

Question 28

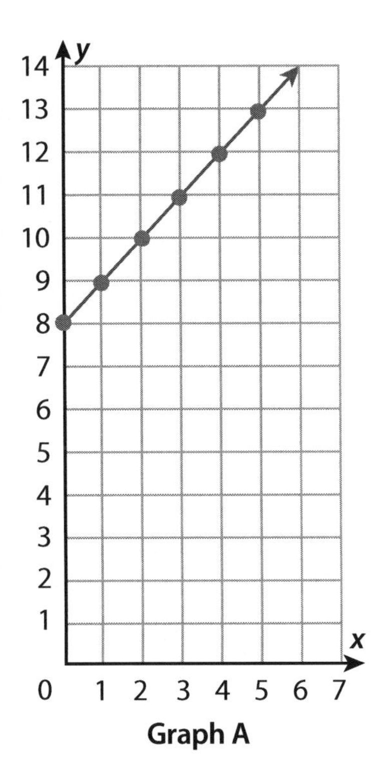

Graph A

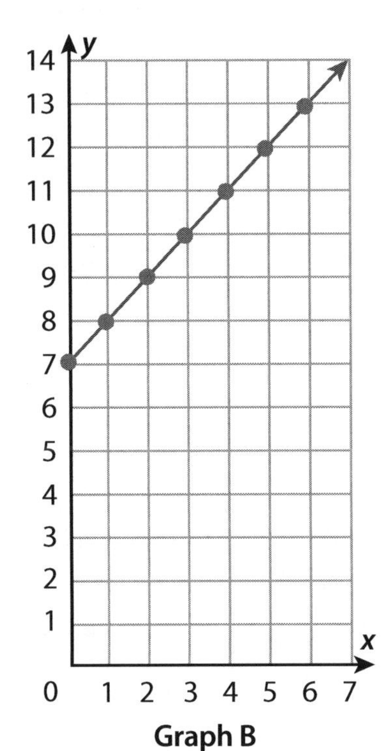

Graph B

Which graph shows this function?

Y = x + 8

A	B
Graph A	Graph B

Question 29

Use approximations to estimate the following sum: $\frac{22.08 \times 8.9}{1.97}$

Answer

Question 30

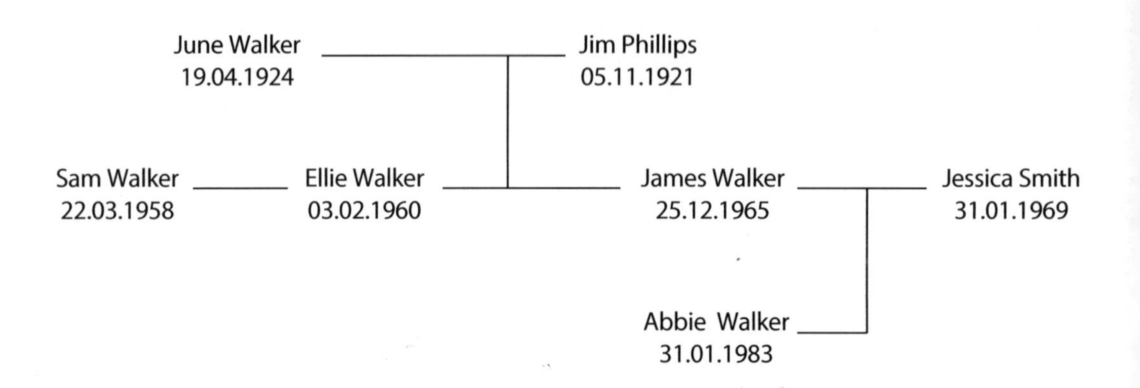

Sam dies before Abbie is born. James was 16 when Sam died. How old was Sam when he died?

A	B	C	D	E
17	23	19	25	28

ANSWERS TO NUMERICAL REASONING – INTERMEDIATE (Section 2)

Q1. D = 6

EXPLANATION = the question may seem tricky at first, but if you notice, the individual number of 6 was cast once in the first 10 attempts, and only once in the first 20 attempts. Therefore, the number 6 could not have turned up from casts 11 – 20.

Q2. C = 5

EXPLANATION = the number 5 only appears 6 times in the first 50 casts, no other number has a lower cast rate at the end of 50 casts, therefore 5 is the number with the least amount of casts in 50 attempts.

Q3. D = 1 and 6

EXPLANATION = the numbers have to occur more than once between 30 and 40. Only the numbers 1 and 6 do this, therefore this would be the correct answer.

Q4. B = 5 : 3

EXPLANATION = total number of 3's cast = 10, total number of 5's cast = 6. The ratio would be 10:6, however the question asks us to put the ratio in its simplest form. Therefore you would need to find a number that goes into both 10 and 6, which would be 2.

So, 10 ÷ 2 = 5 and 6 ÷ 2 = 3. Therefore, the answer would be = 5 : 3.

Q5. D = 676

EXPLANATION = 50 + 81 + 96 + 81 + 86 + 50 + 91 + 65 + 76 = 676.

Q6. E = none

EXPLANATION = 90 + 85 + 58 + 70 + 85 = 388. None of the answers match, so therefore the answer must be 'none'.

Q7. B = 75.2

EXPLANATION = average marks across all subjects = 83 + 73 + 93 + 73 + 54 = 376. Total average mark = 376 ÷ 5 = 75.2.

Q8. D = Elliott

EXPLANATION = Elliott's total if he scored the lowest in all the grade boundaries = 393. Nobody else scored a higher mark.

Q9. D = 39

EXPLANATION = 11 x 36 = 396. 13 x 41 = 533. 533 + 396 = 929 ÷ (11 + 13) = 38.708. To the nearest whole number = 39.

Q10. B = 1.76 hectares

EXPLANATION = 220 x 80 = 17,600 m2. 17,600 ÷ 10,000 = 1.76 hectares.

Q11. C = $594

EXPLANATION = in order to work out the exchange rate, you need to multiply the amount (£450) by the exchange rate for which you are changing into ($).

So, 450 x 1.32 = $594.

Q12. 11.45am

EXPLANATION = 10pm + 11 hours 45 minutes = 09.45. Plus 2 hours time difference (ahead) = 11.45am.

Q13. 0.044

EXPLANATION = in order to work out how to multiply decimals, multiply the numbers normally, ignoring the decimal points. Then put the decimal points back into the answer – remember, it will have as many decimal places as the two original numbers combined.

So, 4 x 11 = 33.

To get 4 from 0.04, it has 2 decimal places. To get 11 from 1.1, it has 1 decimal place. Therefore your answer needs to contain 3 decimal places = 0.044.

Q14. 1 1/8

EXPLANATION = an easy way to remember how to divide fractions is to turn the last fraction upside down, and then multiply.

So, 2/3 becomes 3/2

So, 6/8 x 3/2 = 6 x 3 = 18. 8 x 2 = 16. So, we have the fraction = 18/16

This then needs to be simplified, both numbers can be divided by 2 to make9/8.

Finally, we need to change this fraction into a mixed fraction. 8 goes into 9 once, so that is our number before the fraction. We then know 1 is remaining from (9-8), and the number 8 will remain on the bottom of the fraction to form: 1 1/8.

Q15. 24,000 centimetres

EXPLANATION = first, we need to convert 8 minutes into seconds = 8 x 60 = 480.

Next, we need to multiply 480 (seconds) by the speed (0.5) = 480 x 0.5 = 240.

240 = 240 metres, the question specifically asks to convert this into centimetres. (100cm is equivalent to 1 metre). So, 240 x 100 = 24,000.

Q16. 4.05 kg

EXPLANATION = first add up all the percentages = 45 + 21 + 7 = 73. So, this means that 27% of his allowance is spare.

So, to work out how much this is on the basis of 15 kg being the limit = 15 ÷ 100 x 27 = 4.05. So, James has 4.05 kg allowance to spare.

Q17. E = 14 days

EXPLANATION = 0.25 x 3 = 0.75 (a day).

So, 10.5 ÷ 0.75 = 14 days.

Q18. B = 10 cubic units

EXPLANATION = as you will notice, the diagram shows 10 blocks, therefore the volume of the shape will have to be 10 cubic units.

Q19. D = 2 1/4

EXPLANATION = this may seem tricky, but you must remember that '9' is also a fraction. You need to add the 1 underneath it to make 9/1.

So, 9/1x 2/8 = 9 x 2 = 18 and 1 x 8 = 8. This gives us the fraction = 18/8. This can be simplified to 9/4 and as a mixed fraction, is equivalent to = 2 1/4.

Q20. C = 3 x 3

EXPLANATION = you need to work out which answer option includes prime numbers that are also factors of 9. 3 is a prime number, the question is, how do we get from 3 to 9, still using only prime numbers?

3 x 3 = 9 and therefore contains factors of 9 and are still prime numbers.

Q21. B = 100

EXPLANATION = first you need to work out the percentage of the people who walked = 85% of 65 = 65 ÷ 100 x 85 = 55.25.

Now, you need to work out the percentage of the people who biked = 90% of 50 = 50 ÷ 100 x 90 = 45.

So, 55.25 + 45 = 100.25. Rounded to the nearest whole number = 100.

Q22. A = 45 seconds

EXPLANATION first, you need to add up all the totals from race 1 to 4. So, 240 + 220 + 230 + 210 = 900. You then need to find the average, so 900 ÷ 4 = 225.

Harrison run 800 m in 180 seconds in race 5, so the difference in time from race 5 and the average of his previous races = 225 – 180 = 45.

Q23. 3.40 kg per m^3

EXPLANATION = to work out the density, you need to divide the mass by the volume.

So, 9 ÷ 2.65 = 3.396.

To two decimal places = 3.40 kg per m^3.

Q24. 2 g per cm^3

EXPLANATION = to work out the density, you need to divide the mass by the volume.

So, 60 ÷ 30 = 2 g per cm^3.

Q25. 75,600 grams

EXPLANATION = to work out the mass, you need to multiply the density with the volume.

So, 2.8 x 27,000 = 75,600 grams.

Q26. 79.4

EXPLANATION = for paper 1 = the score was 60 out of 80 which accounted for 35% of the total mark. So, 60 ÷ 80 x 35 = 26.25.

Paper 2 = 45 ÷ 60 x 30 = 22.5.

Paper 3 = 35 ÷ 40 x 35 = 30.625.

Finally, you need to add up all these percentage scores. So, 30.625 + 22.5 + 26.25 = 79.375.

To one decimal place = 79.4.

Q27. 0.82

EXPLANATION = 37 ÷ 45 x 100 = 82.222. To the nearest whole number = 82%. To convert the percentage to a decimal, you need to divide by 100.

So, 82 ÷ 100 = 0.82.

Q28. A = graph A

EXPLANATION = the function: Y = X + 8, so if the number is 2 on X axis, the number on the Y axis would be (2 + 8 = 10).

Q29. 99

EXPLANATION = in order to approximate the numbers, you need to work out how to round the numbers up or down to provide a reasonable approximation.

So, 22.08 can be rounded to 22.

8.9 can be rounded to 9.

1.97 can be rounded to 2.

So, to compete the sum = 22 x 9 = 198 ÷ 2 = 99.

Q30. B = 23

EXPLANATION = James was 16 = 1965 + 16 = 1981. So Sam died in 1981, 1981 – 1958 = 23.

Numerical Reasoning – ***ADVANCED***

Our advanced Numerical Reasoning section will provide you with the skills and knowledge you will be expected to demonstrate when taking a paper that contains particularly challenging mathematical questions. The difficulty of the questions will depend on the type of Numerical Reasoning test you take.

In order for you to successfully pass a Numerical Reasoning test, we have done our utmost to ensure you with lots of questions that focus specifically on an advanced level.

In this type of advanced Numerical Reasoning test, you can expect to find questions on the following areas:

- Percentages
- Fractions
- Decimals
- Data Interpretation
- Equations
- Quantitative Data
- Increases / Decreases
- Speed / Distance / Time
- Mean / Mode / Median / Range
- Box and Whisker Plots
- Statistics
- Currency
- Mass / Density / Volume
- Stem and Leaf Diagrams

Whilst we have provided you with an array of questions, your Numerical Reasoning test will be tailored to the job for which you are applying, and so the questions may not be the same, but will test the same skills and knowledge in terms of an advanced level of mathematics.

EXAMPLES of Numerical Reasoning - ***ADVANCED***

Hectares

Work out the area of the shape. Write your answer in hectares.

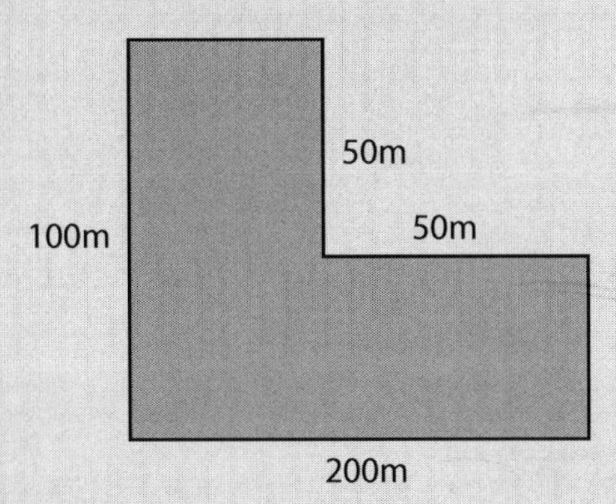

Hectares

The important thing to remember when dealing with hectares, is to use this information as a guideline:

1 hectare = 10,000 m = 2.47 acres

So, if you had measurements in centimetres, you would need to convert them into metres before attempting to convert the measures into hectares.

Example

For the above example, to work out the area of the shape in hectares, you need to work out the area of the shape first.

Step 1 = 100 x 200 = 20,000

Step 2 = 50 x 50 = 2,500

Step 3 = 20,000 – 2,500 = 17,500m^2

Step 4 = to convert 17,500m^2 into hectares, we know 10,000 m^2 = 1 hectare.

So, 17,500 ÷ 10,000 = 1.75 hectares.

Percent Increase

Work out the percentage increase.

Percent Increase

To work out the percentage increase of a set of data, you need to remember this formula:

Percent Increase % = Increase ÷ original number x 100

If your answer is a negative number, then this is a percentage decrease.

Percent Decrease

Work out the percentage decrease.

Percent Decrease

To work out the percentage decrease of a set of data, you need to remember this formula:

Percent Decrease % = Decrease ÷ original number x 100

If your answer is a negative number, then this is a percentage increase.

Velocity Graphs

How much greater is the acceleration of Car B than the acceleration of Car A?

You need to remember this formula:

Acceleration (m/s2) = Change in velocity (m/s2) ÷ Change in time (s)

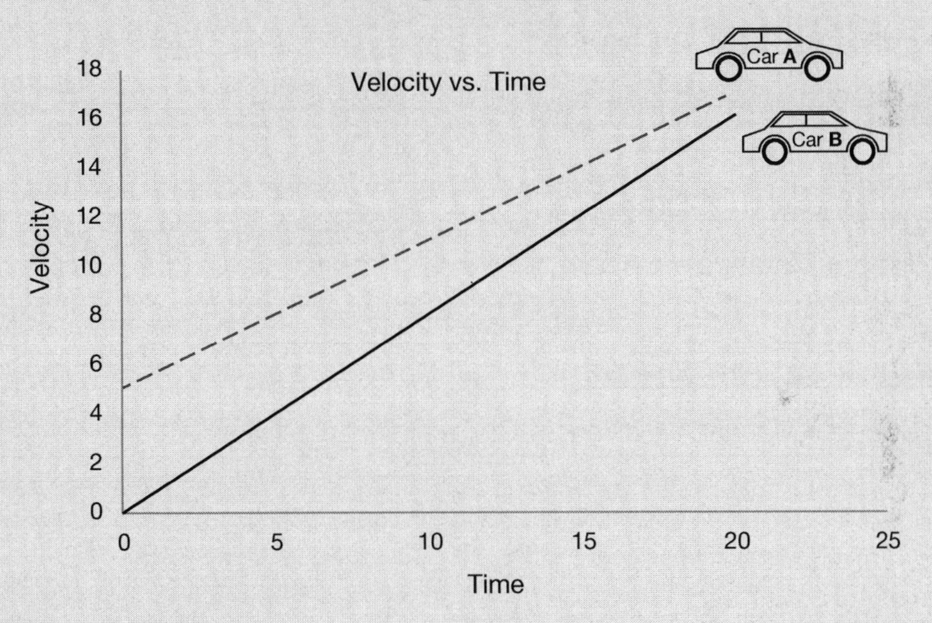

<u>How to work it out</u>

Step 1 = Car A = change in velocity = 16 – 5 = 11

$11 \div 20 = 0.55$

Step 2 = Car B = change in velocity = 15 – 0 = 15

$15 \div 20 = 0.75$

Step 3 = difference from Car B to Car A = 0.75 – 0.55 = 0.2

Speed / Distance / Time

Speed, distance and time.

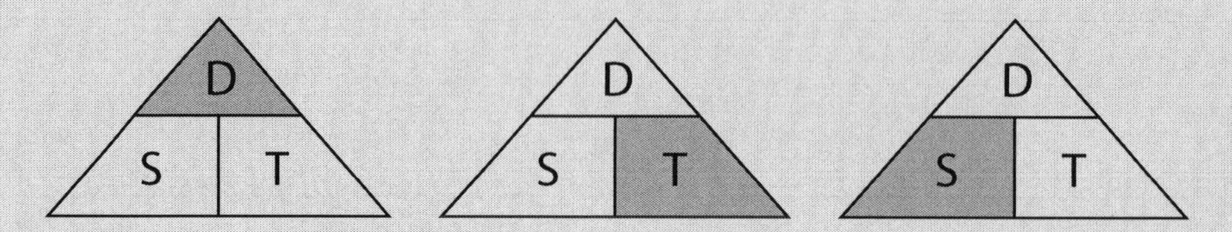

To work out the Distance:

- Distance = Speed x Time

To work out the Time:

- Time = Distance ÷ Speed

To work out the Speed:

- Speed = Distance ÷ Time

Equation Correspondence

A square field, S, has an area greater than 5000 m^2. Its length is decreased by 20m and its width also decreases by 20m to give a rectangular field, R. Which one of the following is true?

A. Perimeter R = area R and perimeter S > area S
B. Area R < area S and perimeter S = perimeter R
C. Area S > area R and perimeter S > perimeter R
D. Area S = area R and perimeter S < perimeter R
E. Area S < area R and perimeter S < perimeter R

Answer = C

How to work it out

- You need to know what '<' '>' and '=' mean in order to work these out.
- '<' means a small number followed by a large number.
- '>' means a large number followed by a small number.
- '=' means the two numbers are equal.

Powers

Solve 10 x 4^{11}.

- This does not mean 4 x 11.
- 4^{11} means = 4 x 4 x 4 x 4 x 4 x 4 x 4 x 4 x 4 x 4 x 4 = 4,194,304

Gradients

Negative gradients

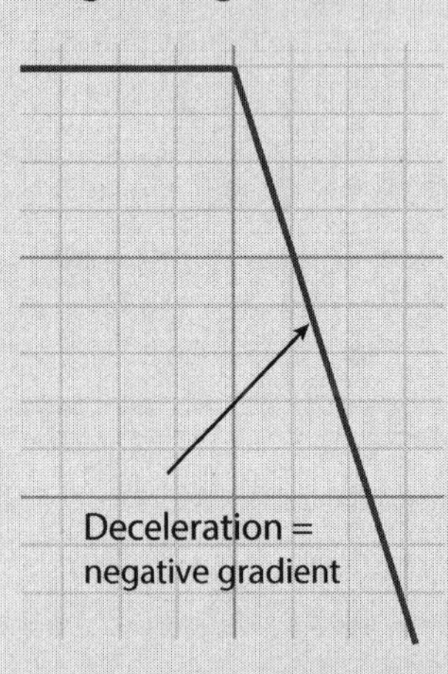

The negative gradient of a graph is the line that is going downwards (not upwards), if the gradient was going upwards, this would be a positive gradient.

Exchange Rate

- If you had to exchange £200 into euros which had the exchange rate of 1.1.46, you would multiply how much you want to exchange (£200) by the exchange rate (1.46).
- So, 200 x 1.46 = 292 euros.

Areas / Perimeters

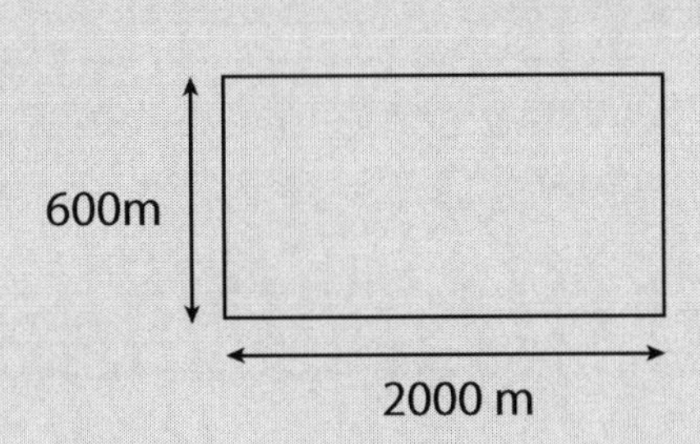

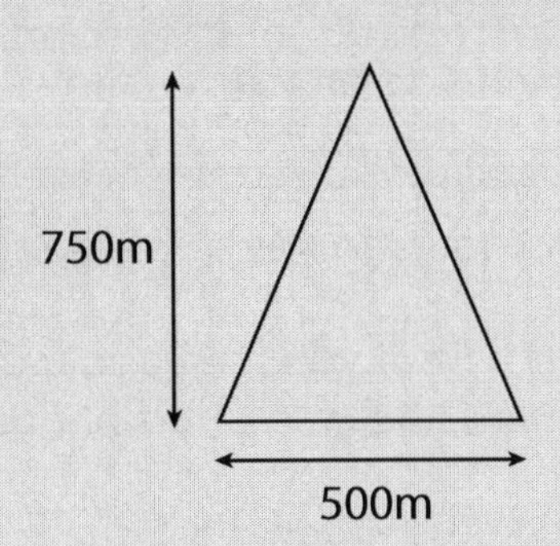

Area of squares/ rectangles

Base x height

- 2,000 x 600 = 1,200,000m^2

Area of triangles

½ base x height

- ½ x 500 x 750 = 187,500

Perimeter of squares / rectangles

Add all the sizes of each side.

- 2,000 + 2,000 + 600 + 600 = 5200

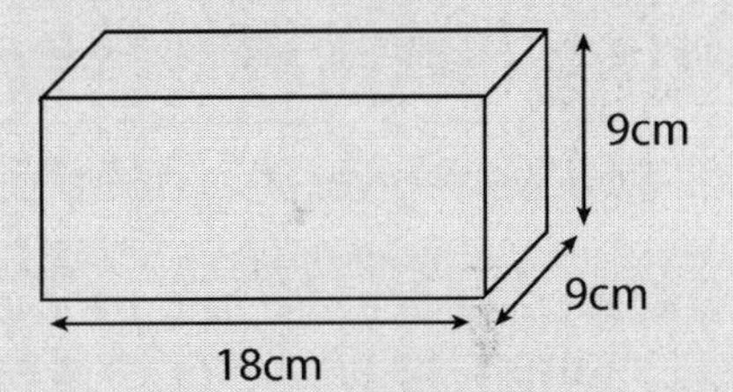

Volume

Length x base x height

- 18 x 9 x 9 = 1458

Compound Interest

Compound Interest

If the question is asking you to work out the compound interest, this means that "interest is added on to the interest".

For example, if you financed a car and had to pay 4.6% interest per year for 3 years, including compound interest, based on the rate of £560.

- 560 ÷ 100 x 104.6 = 585.76
- You then use this sum (585.76) to work out the next interest rate.
- 585.76 ÷ 100 x 104.6 = 612.70.
- For the third year, 612.70 ÷ 100 x 104.6 = 640.88.

Water Level Rises

If you were given the question:

30 men take a dip in a swimming pool 40m long and 30m broad. If the average displacement of water by a man is 4m^3, then what will the rise in the water level be?

How to work it out:

- Total volume of water displaced = 4m^3 x 30 (men) = 120m^3.
- Rise in water level = 120 ÷ (40 x 30) = 120 ÷ 1200 = 0.1m x 100 = 10cm.
- So, the rise in water level would be 10cm.

Numerical Reasoning – ***ADVANCED***

(Section 1)

Question 1

The diagram below shows the plan of a building site. All angles are right angles.

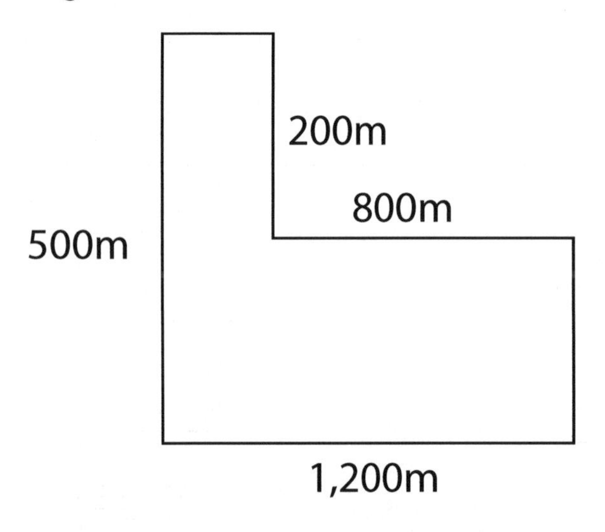

What is the area of the building site? Give your answer in hectares.

1 hectare = 10,000m² = 2.47 acres.

A	B	C	D
60 hectares	40 hectares	44 hectares	4.4 hectares

Question 2

The following graph shows the velocity of two cars at different times.

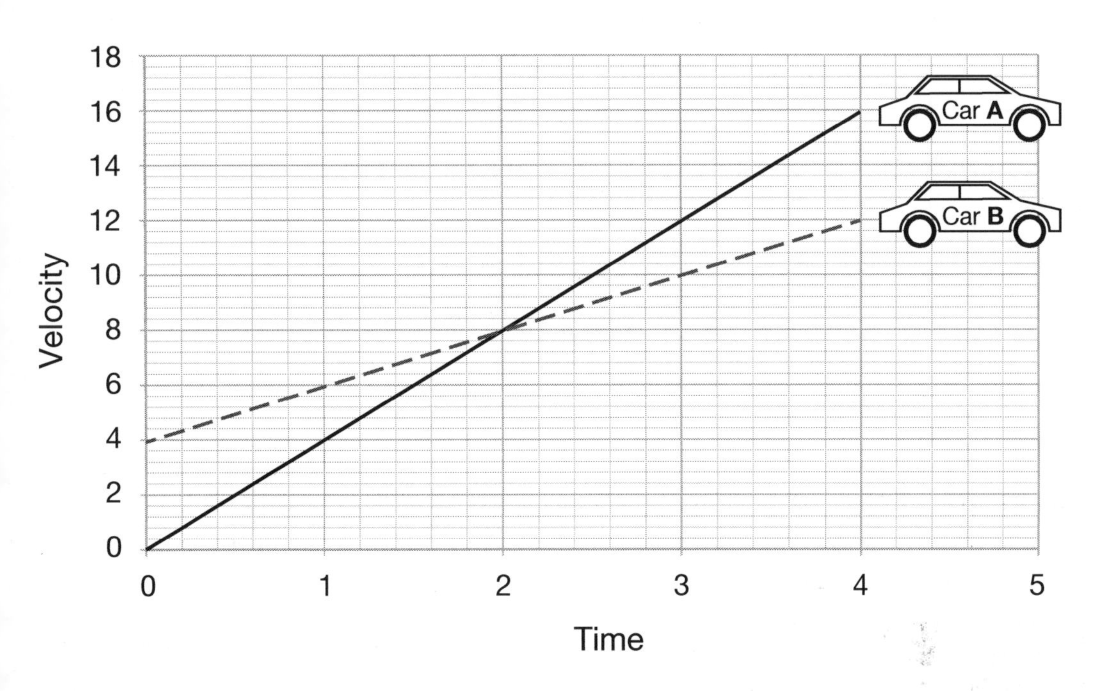

How much greater is the acceleration of Car A than the acceleration of Car B?

Acceleration (m/s2) = Change in velocity (m/s2) ÷ Change in time (s)

A	B	C	D
2 m/s2	4 m/s2	6 m/s2	8 m/s2

Question 3

Government spending on "Education services" and "Health services" was 56.3 billion pounds and 106.7 billion respectively for the year 2009-2010. In the same year, the Government spending on "Debt Interests" was 22.22% of the spending on "Education services". The spending on 'Education Services', "Health services" and "Debt Interests" constituted 50% of the total spending by the Government.

What was the Government's approximate total spending for the year 2009-2010?

A	B	C	D
551 billion pounds	615 billion pounds	351 billion pounds	435 billion pounds

Question 4

The Siberian tiger population in Country A is 60% of the Siberian tiger population in Country B. The population of Siberian tigers in Country C is 50% of that in the Country A.

If the Siberian tiger population in Country C is 420, what is the Siberian tiger population in Country B?

A	B	C	D
1,400	1,200	1,000	1,600

Question 5

The graph shows respondent's answers when asked what their most frequent form of technological communication was.

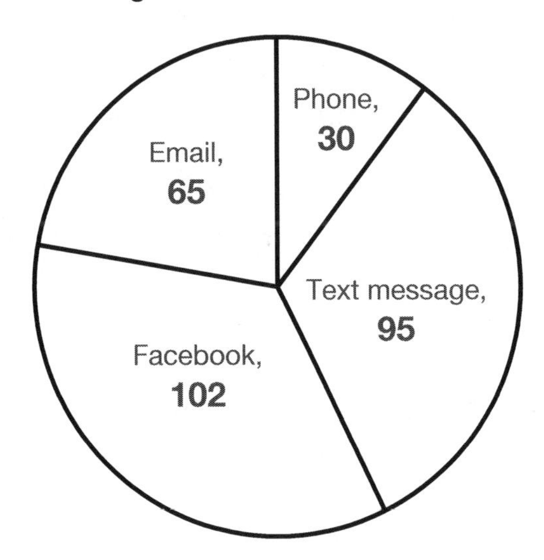

Among the respondent's, 75% of all respondents said that it was easier to get in touch with someone through technological communications based on convenience. How many people said that it was easier to use technological communications based on convenience?

A	B	C	D
212	119	227	219

Question 6

In a survey, people had to choose either A, B, C or D.

The percentages for A, C and D are shown below.

A	B	C	D
25%		30%	15%

320 people chose A. How many people chose option B?

A	B	C	D
215	384	429	502

Question 7

If Mike's train leaves at 8.30am rather than 8am and maintained an average speed of 48.3 km/h throughout the journey without any delays, what time would Mike arrive home? 100 miles = 161 km

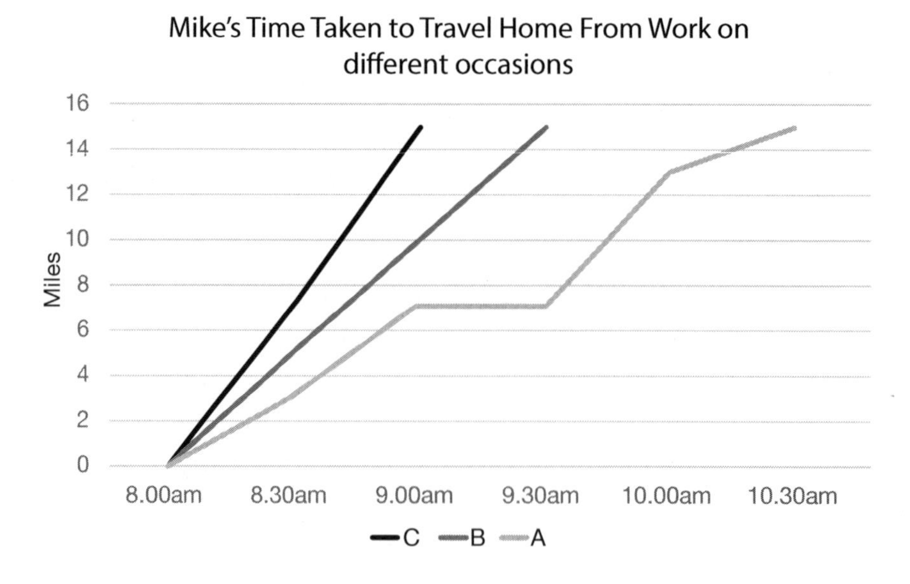

Answer

Question 8

Solve $5x - 2 = x + 16$

Answer

Question 9

STORE	PERCENT CHANGE FROM 2011 TO 2012	PERCENT CHANGE FROM 2012 TO 2013
U	18	-10
V	17	-7
W	16	6
X	20	-5
Y	-15	-8

If the dollar amount of sales at Store W was $456,250 for 2011, what was the dollar amount of sales at that store for 2013?

Answer

Question 10

Here is a projected and estimated number of passengers.

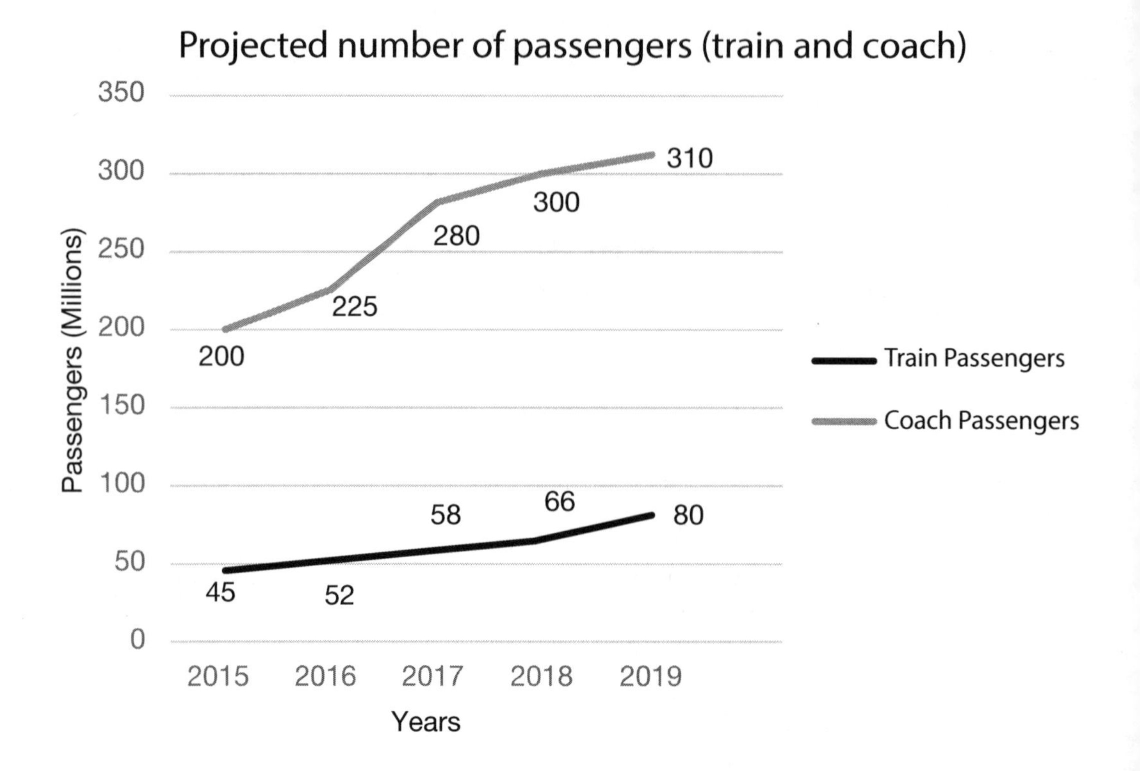

What is the projected ratio of coach passengers to train passengers in 2019?

A	B	C	D	E
5 : 1	3 : 1	31 : 8	27 : 9	Cannot say

Question 11

Below is a table showing today's exchange rate for dollars.

	Closing Point	Today's High	Today's Low	Closing Point change from yesterday
Sterling	35.20	37.16	33.19	+0.33
Euro	20.22	22.04	19.60	-0.30
Yen	0.24	0.28	0.23	+0.11

How many Euros would you get for £300 sterling yesterday, according to the closing point exchange rate (rounded to the nearest 10)?

Answer

Question 12

Imagine we have measured a distance of 7cm on a map. The map has a scale of 1:4000m2. Work out the distance in real life. Your answer should be in hectares.

1 hectare = 10000m2 = 2.47 acres

A	B	C	D	E
2.8 hectares	28 hectares	8 hectares	1 hectare	20 hectares

Question 13

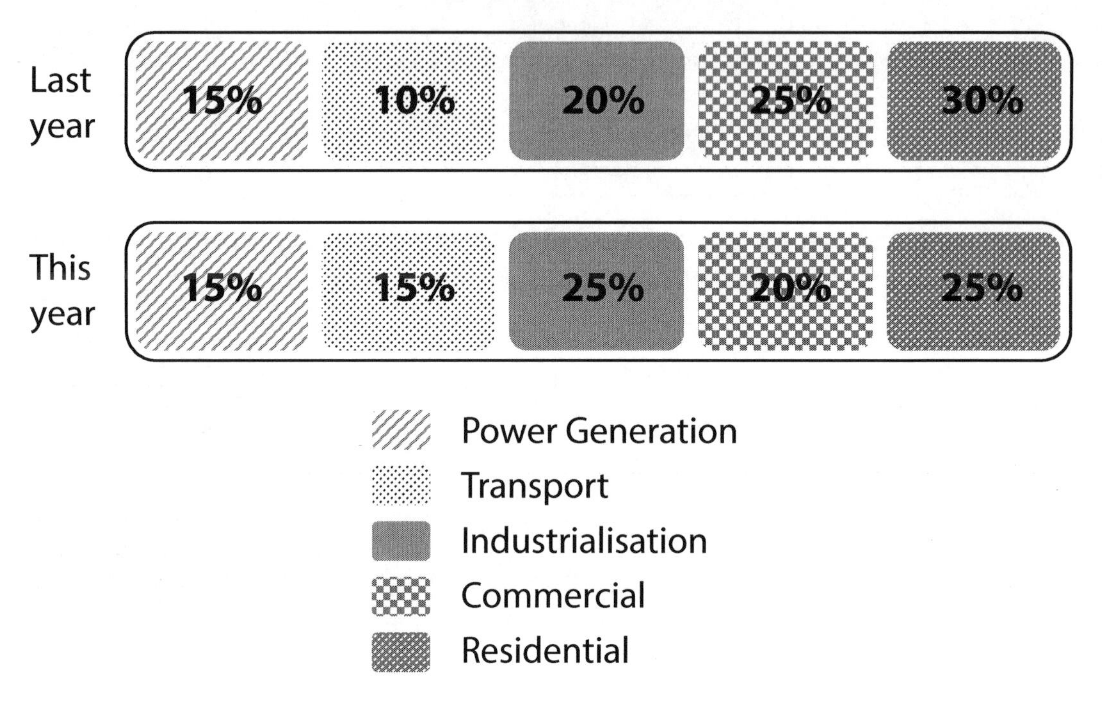

If transport emitted 6 million tons this year, and industrial emissions are the same as last year, what were the commercial emissions last year?

A	B	C	D	E
1 million ton	4 million tons	3 million tons	5 million tons	8 million tons

Question 14

Below is a table representing the income of industries in billions of pounds over a five year period.

	Year 1	Year 2	Year 3	Year 4	Year 5
Financing	65	82	93	100	112
Telecommunications	18	21	27	34	38
Engineering	37	58	60	64	68
Agriculture	26	26	30	30	55
Media	59	60	72	78	75
Manufacturing	33	38	41	30	27
Transportation	48	48	49	56	60

Which industry had the largest increase in amount of income between Year 2 and Year 3?

A	B	C	D	E
Financing	Transportation	Media	Agriculture	Engineering

Question 15

Below is a table of the total staff at Company A (Staff Distribution).

	HR	Sales	Finance	Media	Distribution	TOTAL(%)
Year 1	21	8	19	32	20	100
Year 2	28	11	17	28	16	100
Year 3	16	21	19	26	18	100
Year 4	14	29	21	14	22	100
Year 5	4	9	25	38	24	100
Year 6	20	27	25	12	16	100

In Year 4, there were 406 people employed in Finance. How many people in total were employed in Year 4 in the department of Sales? To the nearest whole person.

Answer [ ]

Question 16

The area of the District of Columbia is 177km². How many hectares is that?

1 hectare = 10,000 m² = 2.47 acres.

A	B	C	D	E
177 hectares	107 hectares	7 hectares	47 hectares	17.7 hectares

Question 17

Work out which equation corresponds Shape A with Shape B.

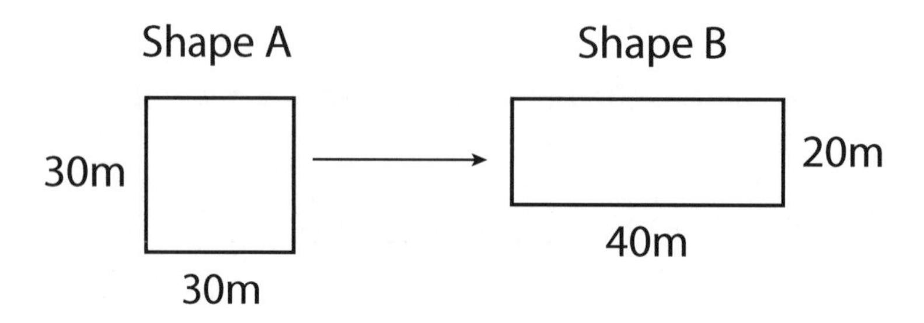

A. Area A < area B and perimeter A = perimeter B

B. Area A = Area B and perimeter A = perimeter B

C. Area A > area B and perimeter A > perimeter B

D. Area A > area B and perimeter A = perimeter B

E. Area A < area B and perimeter A < perimeter B

Answer [ ]

Question 18

Below is a bar chart demonstrating the prices and number of package holidays in the year 2014.

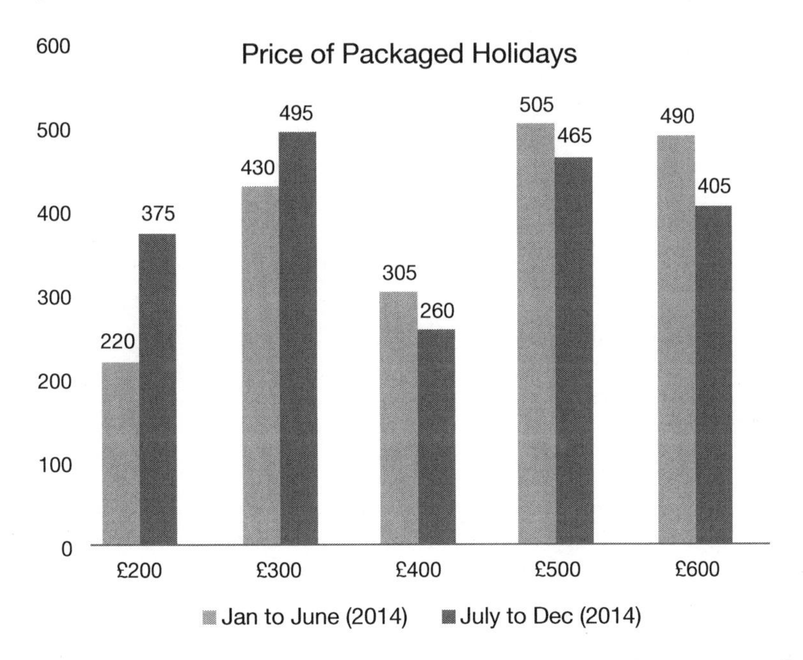

Each sales agent who manages to sell a package holiday receives a paid commission of 3.5% of the total cost of the holiday. How much commission is paid out in total to all sales agents for the year 2014?

A	B	C	D
£49,995.50	£57,557.50	£41,395.95	£61,225.75

Question 19

Using the above bar chart, what is the mean number of holidays sold in 2014 from July to December?

A	B	C	D
1,000	200	800	400

Question 20

In July, Ryan worked a total of 40 hours, in August he worked 46.5 hours – by what percentage did Ryan's working hours increase in August?

A	B	C	D
16.25%	165%	1.625%	25%

Question 21

There were 17 million families in the UK in 2006.

The total income of families in the UK was £5.60 × 10^{11}.

What was the mean income per family?

Give your answer to the nearest thousand.

A	B	C	D
£30,000	£28,000	£33,000	£33,900

Question 22

There were 17 million families in the UK in 2006.

The mean number of children per family was 1.8.

How many children were there in the UK?

A	B	C	D
30.6 million	306 million	3.6 million	36 million

Question 23

The following graph shows the marks of a student who has taken their Maths GCSE four times.

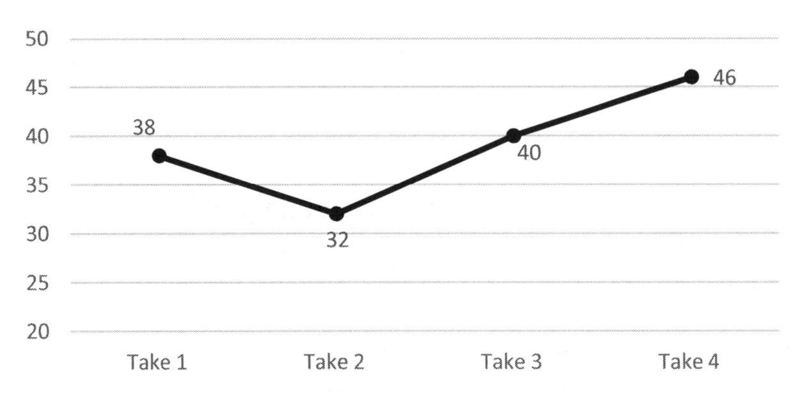

In the student's final take (take 5), they scored 68 and therefore passed their GCSE. How many more marks did the student receive in this exam, than the average mark of their previous attempts at the exam?

A	B	C	D
20	29	31	35

Question 24

Below is a table representing the number of deliveries in March 2014.

Delivery Area	Truck A	Truck B	Truck C	Truck D	Truck E
Cardiff	315	255	354	269	466
Manchester	759	436	157	357	143
Brighton	135	764	125	456	421
Dover	355	874	477	258	465
Portsmouth	551	668	567	776	904

It is expected that there will be a 35% increase in demand for deliveries in March the following year, and a new truck will need to join the company.

If the new truck covers all the new deliveries, how many deliveries in total will the new truck have to deliver next March? Rounded up to the nearest whole number.

A	B	C	D
3,250	3,350	3,790	3,950

Question 25

A business spending on "HMRC" and "Salaries" was 35 million pounds and 40 million respectively for the year 2012-2013. In the same year, the business spending on "Pensions" is 20% of the total spending of "HMRC". The spending on "HMRC", "Salaries" and "Pensions" constituted 50% of the spending expenditures for the business in the year 2012-2013.

What is the approximate total spending for the business in the year 2012-2013?

A	B	C	D
152 million pounds	164 million pounds	150 million pounds	172 million pounds

Question 26

The following pie chart demonstrates the approximate population data for China, India and the U.S.A.

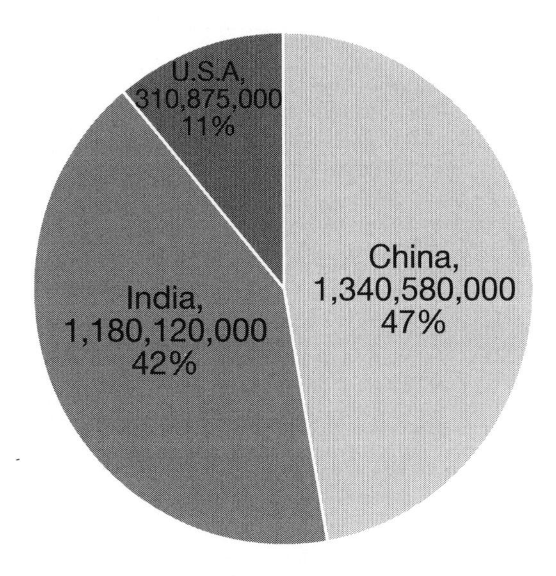

If India's population was once the same as China's, what is the percentage decrease in India's population from then to now? Rounded to two decimal places.

A	B	C	D
11.969%	11.70%	11.9%	11.97%

Question 27

Using the above pie chart, India's unemployed population is approximately half the unemployed population of China, and 116,044,032 more than the U.S.A's unemployed population of 16,285,800. What is the approximate unemployed population of China?

A	B	C	D
64,646,991	66,164,916	54,125,875	67,482,165

Question 28

Below is a velocity-time graph.

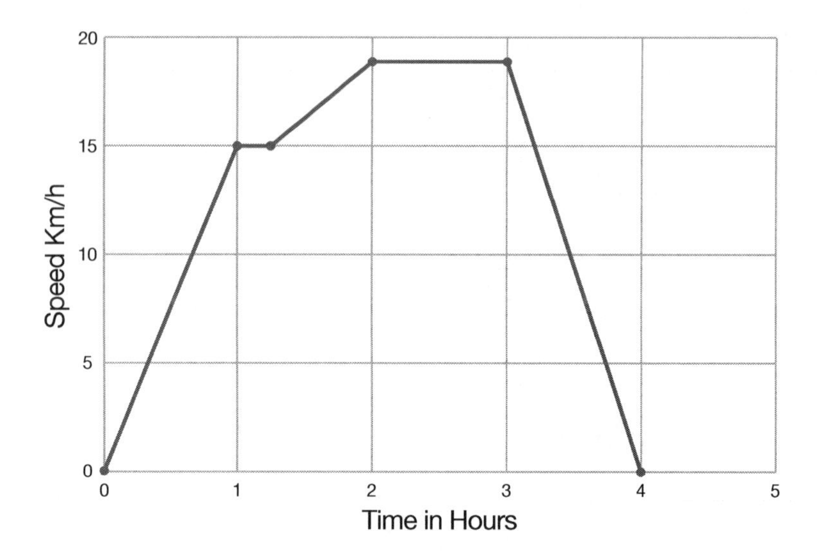

How many times was the acceleration zero?

Answer

Question 29

Using the above velocity-time graph, what was the distance travelled in the first half hour?

Answer

Question 30

Using the above velocity-time graph, what was the deceleration shown by the graph?

A	B	C	D
15 km/h	20 km/h	19 km/h	9 km/h

ANSWERS TO NUMERICAL REASONING – ADVANCED (Section 1)

Q1. C = 44 hectares

EXPLANATION = Work out the area of the whole shape: 1200 x 500 = 600,000

Work out the area of the missing rectangle (to make a complete rectangle): 800 x 200 = 160,000

So, 600,000 – 160,000 = 440,000m^2.

440,000m^2 in hectares = 440,000 ÷ 10,000 = 44 hectares.

Q2. A = 2 m/s2

EXPLANATION = (y final – y initial) ÷ (x final – x initial).

Car A = (16-0) ÷ (4-0) = 16 ÷ 4 = 4.

Car B = (12-4) ÷ (4-0) = 8 ÷ 4 = 2.

So, the difference between car A and car B is 2 m/s2.

Q3. C = 351 billion pounds

EXPLANATION = Education services = 56.3 billion pounds and Health services = 106.7 billion pounds.

22.22% of 56.3 = 56.3 ÷ 100 x 22.22 = 12.50986 (Round up = 12.51).

The total of Education, Health and Debt Interests = 175.51 billion pounds.

The total Government spending = 175.51 x 100 ÷ 50 = 351.02.

So, the approximate total = 351 billion pounds.

Q4. A = 1,400

EXPLANATION = Siberian tiger population in Country C is 50% of that in Country A.

If country C is 420, Country A = 420 x 100 ÷ 50 = 840.

So, if Country A = 840 and is 60% of the population in Country B, Country B = 840 x 100 ÷ 60 = 1,400.

Q5. D = 219

EXPLANATION = first, you need to add up all the respondents who took part in the survey: 65 + 30 + 95 + 102 = 292.

Next, you need to work out 75% of 292: 292 ÷ 100 x 75 = 219.

So, 219 respondents said they use certain technological communications for convenience.

Q6. B = 384

EXPLANATION, first you need to work out how many people took the survey in total: 320 (people who chose A) x 100 ÷ 25(%) = 1280.

Next, you need to work out the missing percentage for B. 100(%) – 25(%) – 30(%) – 15(%) = 30(%).

So, B is going to be 30% of the total: 1280 ÷ 100 x 30 = 384.

Q7. 9.00am

EXPLANATION = an average speed of 48.3 km/h needs to be worked out in relation to miles per hour.

X miles x 161 km = 100 miles x 48.3 = 4830.

4830 miles ÷ 161 km = 30.

That means that Mike is travelling 30 miles in approximately 1 hour.

His house is 15 miles away from work, so 1 hour needs to be halved = 30 minutes.

So, if Mike left work at 8.30am, he would arrive home at 9.00am.

Q8. 4.5

EXPLANATION = the two on the left side of the equals sign, needs to be moved over to the right side of the equation (so the numbers are on one side, and the algebra is on the other). When moving numbers over, you have to do the opposite to what it says. So, the equation says '-2'; so we need to add 2 to 16 = 18. The 'x' needs to be moved to the left side of the equals sign, so '+x' will become '-x'. So, 5x – 1x = 4x = 18.

4x = 18.

18 ÷ 4 = 4.5.

Q9. $561,005

EXPLANATION = for Store W, in 2011 = $456,250. In order to get from 2011 to 2012, we see a 16% increase. So, 456,250 ÷ 100 x 116(%) = 529,250.

To get from 2012 to 2013, we see a 6% increase. So, 529,250 ÷ 100 x 106 = 561,005.

So, the store amount of sales for Store W in 2013 is $561,005.

Q10. C = 31 : 8

EXPLANATION = the number of coach passengers in 2019 = 310.

The number of train passengers in 2019 = 80.

So, 310 : 80 is our ratio, however this can be simplified.

310 : 80 = both can be divided by 2 = 155 : 40.

155 : 40 = both can be divided by 5 = 31 : 8.

This is in its simplest form.

Q11. 510 euros

EXPLANATION = you need to build a relationship between Euros and Sterling in order to work out how much £300 is worth in Euros.

Yesterday's closing point for Euro's = 20.22 + 0.30 = $20.52. So, $20.52 is worth 1 euro.

Yesterday's closing point for Sterling = 35.20 – 0.33 = $34.87. So, $34.87 is worth £1.

£300 worth in dollars = 34.87 x 300 = £10,461.

You now need to work out how many Euros £10,461 is worth. So, 10,461 ÷ 20.52 = 509.79. Now, you're almost there. The question asks for it to be rounded to the nearest 10 = 510.

Q12. A = 2.8 hectares

EXPLANATION = 7 (cm) x 4000 (m2) = 28,000 (m2).

So, 28,000 ÷ 10,000 = 2 hectares 800 metres.

Q13. D = 5 million tons

EXPLANATION = First, we notice that the ratio of transport to industrial is 15:25, which is 3:5, which is 6:10, so emissions by industry this year were 10 million tons. This is the same as last year, and last year commercial emissions were half industrial emissions, so commercial emissions were 5 million tons.

Q14. C = Media

EXPLANATION = you need to work out the difference for each industry from Year 2 to Year 3.

Financing = increase = 11

Telecommunications = increase = 6

Engineering = increase = 2

Agriculture = increase = 4

Media = increase = 12

Manufacturing = increase = 3

Transportation = increase = 1

So, the largest increase between Year 2 and Year 3 was in the Media industry.

Q15. 560

EXPLANATION = in order to work out the number of people working in Sales in Year 4, you need to work out the total number of employees in that year.

So, 406 (number of people employed in Finance) x 100 ÷ 21 (percentage of Finance) = 1933.333. To the nearest whole person = 1933.

So, 1933 ÷ 100 x 29 (number of employees in Sales) = 560.57. To the nearest whole person = 560.

Q16. E = 17.7 hectares

EXPLANATION = you need to convert 177 km into metres in order to work out the hectares. 177 kilometres x 1,000 = 177,000.

So, 177,000 ÷ 10,000 = 17.7 hectares.

Q17. D = Area A > area B and perimeter A = perimeter B

EXPLANATION = area of Shape A = 30 x 30 = 900m².

Area of Shape B = 40 x 20 = 800m².

So, the area of Shape A is larger than the area of Shape B and so needs (>) sign.

Perimeter of Shape A = 30 + 30 + 30 + 30 = 120

Perimeter of Shape B = 40 + 40 + 20 + 20 = 120

So, the perimeter of Shape A is exactly the same as the Perimeter of Shape B and so needs an (=) sign.

Q18. B = £57,557.50

EXPLANATION = firstly you need to work out each commission for the price of each holiday.

£200 holiday = 3.5% of 200 = 200 ÷ 100 x 3.5 = £7 x 595 (total number of holidays sold for £200) = 4,165.

£300 holiday = 3.5% of 300 = 300 ÷ 100 x 3.5 = £10.50 x 925 = 9712.50.

£400 holiday = 3.5% of 400 = 400 ÷ 100 x 3.5 = £14 x 565 = 7910.

£500 holiday = 3.5% of 500 = 500 ÷ 100 x 3.5 = £17.50 x 970 = 16,975.

£600 holiday = 3.5% of 600 = 600 ÷ 100 x 3.5 = 21 x 895 = 18,795.

So, add up all these totals = 4,165 + 9712.50 + 7910 + 16,975 + 18,795 = £57,557.50.

Q19. D = 400

EXPLANATION = add up all of the holidays from July to December and then divide it by the number of different holidays costs (5).

So, 375 + 495 + 260 + 465 + 405 = 2,000. So, 2,000 ÷ 5 = 400.

So, the mean number of holidays from July to December is 400.

Q20. A = 16.25%

EXPLANATION = To tackle this problem first we calculate the difference in hours between the new and old numbers. 46.5 - 40 hours = 6.5 hours. We can see that Ryan worked 6.5 hours more in August than he did in July – this is his increase.

To work out the increase as a percentage it is now necessary to divide the increase by the original (January) number: 6.5 ÷ 40 = 0.1625

Finally, to get the percentage we multiply the answer by 100. This simply means moving the decimal place two columns to the right.

0.1625 × 100 = 16.25.

Ryan therefore worked 16.25% more hours in August than he did in July.

Q21. C = £33,000

EXPLANATION = (5.6 x 10^{11}) ÷ 17,000,000 = 32,941. (Please note that 10^{11} does NOT mean 10 x 11, it means 10 x 10 x 10 x 10 x 10 x 10 x 10 x 10 x 10 x 10 x 10).

To the nearest thousand = £33,000.

Q22. A = 30.6 million

EXPLANATION = 17,000,000 (number of families in the UK) x 1.8 (mean number of children per family) = 30,600,000.

Q23. B = 29

EXPLANATION = average mark on previous marks = 38 + 32 + 40 + 46 = 156 ÷ 4 = 39.

Difference from his pass mark (68) and his average mark from the previous takes (39) = 68 – 39 = 29.

Q24. D = 3950

EXPLANATION = add up all the numbers in the table = 11,287.

To work out a 35% increase, 11,287 ÷ 100 x 135 = 15,237.45. To the nearest whole number = 15,237.

So, to work out how many deliveries the new truck will be making = 15,237 – 11,287 = 3,950

Q25. B = 164 Million pounds

EXPLANATION = HMRC services = 35 million pounds and Salaries = 40 million pounds.

20% of 35 = 7 million pounds.

The total of HMRC, Salaries and Pensions = 82 million pounds.

The total Government spending = 82 x 100 ÷ 50 = 164 million pounds.

Q26. D = 11.97%

EXPLANATION = to work out the percentage decrease = original number – new number. Decrease ÷ original number x 100.

So, 1,340,580,000 – 1,180,120,000 = 160,460,000.

160,460,000 ÷ 1,340,580,000 x 100 = 11.969.

To two decimal points = 11.97% decrease.

Q27. B = 66,164,916

EXPLANATION = the information reveals that India's unemployed population is 116,044,032 more than the U.S.A's which is 16,285,800.

To work out the unemployed population of India = U.S.A's unemployed population + 116,044,032.

So, China's unemployed population = 116,044,032 + 16,285,800 = 132,329,832 ÷ 2 = 66,164,916.

Q28. 2

EXPLANATION = each flat part of the graph shows zero acceleration. Therefore, the graph shows two incidents where the acceleration is zero and so the correct answer is 2.

Q29. 7.5 km

EXPLANATION = area under graph = distance travelled = ½ x base x height = ½ x 1 x 15 = 7.5 km.

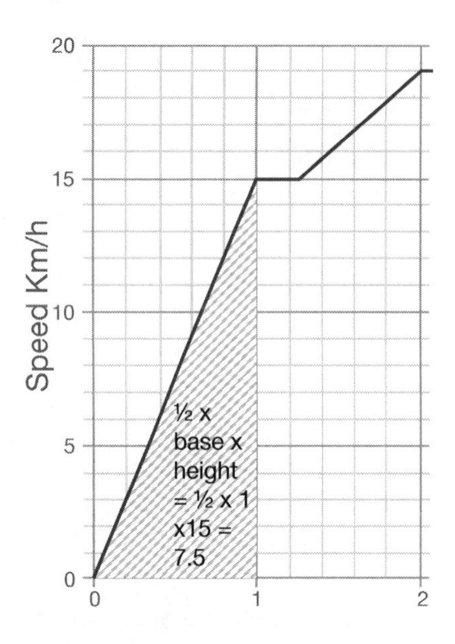

Q30. C = 19 km/h

EXPLANATION = Deceleration is the negative gradient. = Change in Y ÷ Change in X = 19 km/h ÷ 1 hour = 19 km/h.

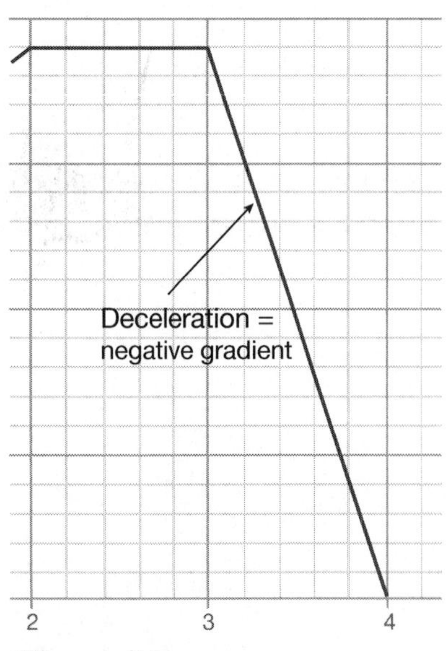

Numerical Reasoning – ***ADVANCED***

(Section 2)

Question 1

Below is a table showing the population data for Countries A, B, C and D in the year 2012.

	Population at the beginning of the year (millions)	Unemployment rate (%)	Annual Population Growth Rate (%)
Country A	6	8.5	0.5
Country B	10	8.2	0.2
Country C	10	7.4	0.8
Country D	12	6.5	0.8

If the annual population growth rate for Country B remains constant, how many years are required for the population of Country B to increase by 120,600?

A	B	C	D
2 years	5 years	6 years	9 years

Question 2

Using the above table, if both the annual population growth rate and unemployment rate remain the same for all four countries, which country will have the greatest unemployment rate by 2014?

A	B	C	D	E
Country A	Country B	Country C	Country D	Cannot be determined

Question 3

Mineral water is classified on the basis of the amount of dissolved solid materials it contains. The chart shows the codes of different levels of total dissolved solids (TDS) and the number of mineral water bottles for each code sold at a store.

MINERAL WATER BOTTLES		
Code	**TDS (mg/l)**	**Number of bottles**
TDS 1	Less than 50	52
TDS 2	Greater than or equal to 50 but less than 500	85
TDS 3	Greater than or equal to 500 but less than 1,500	65
TDS 4	Greater than or equal to 1,500	50

What fraction of the total number of bottles sold at the store with TDS greater than or equal to 50 mg/l, have the code TDS 4?

A	B	C	D
1/2	2/3	3/4	1/4

Question 4

The following information is displayed at the Bank of England.

Currency	Sell at:	Buy at:
Australian Dollars	1.52	1.84
Canadian Dollars	1.54	1.78
Turkish Lira	2.45	3.28
US Dollars	1.48	1.62
Euros	1.16	1.32

How much will 200 Euros cost a customer in Pounds Sterling?

A	B	C	D
£172.40	£232.00	£151.52	£264.00

Question 5

The Earth travels around the sun once a year.

The average distance of the Earth from the Sun is 1.5 x 10^{11}m.

Work out the distance between the earth and the sun. Write your answer in kilometres.

A	B	C	D
150,000,000 km	150,000,000,000 km	1,500 km	1,500,000 metres

Question 6

The population of the world in 1960 was 3040 million. In 1975, it was 4090 million. It is considered that the population grows by a constant percentage each year.

What constant annual percentage growth rate from 1960 to 1975 would result in the population increasing from 3040 million to 4090 million?

A	B	C	D
5 %	8 %	2 %	3 %

Question 7

Here is a scatter graph showing the distance and speed of a cheetah's movements.

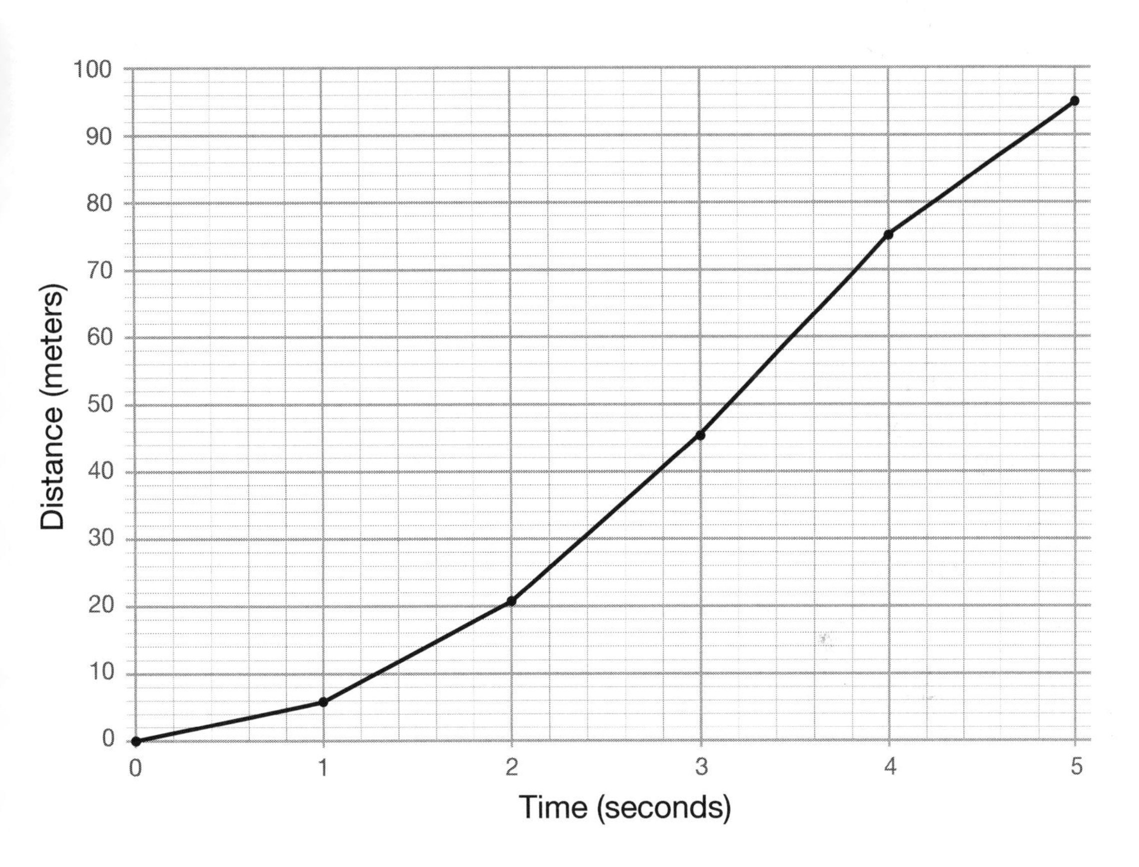

Estimate the average distance the cheetah runs in the first 3 seconds.

Answer

Question 8

The graph below shows the distance travelled by an animal.

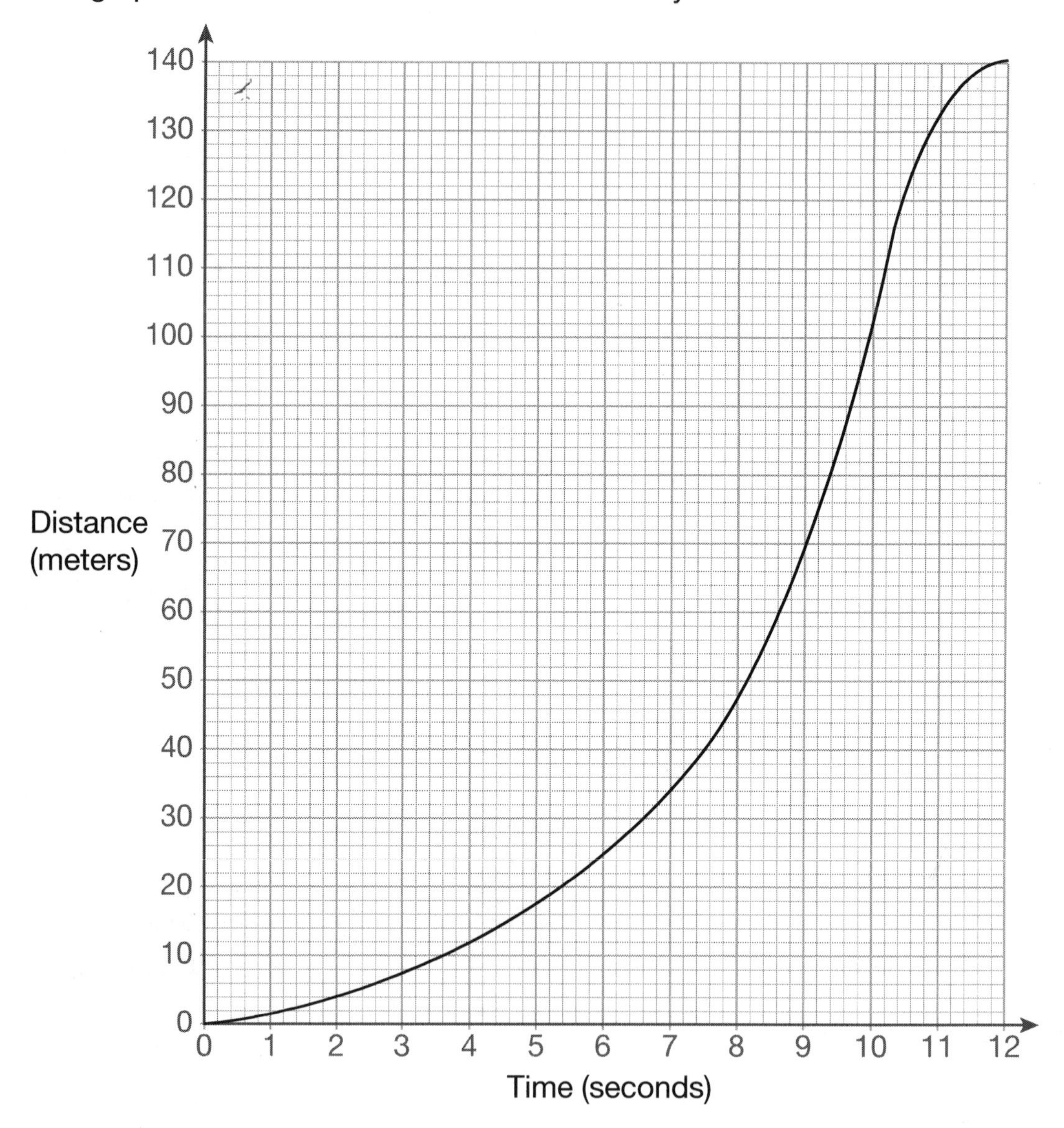

Estimate the speed of the animal at 10 seconds.

A	B	C	D
12.5 metres per second	10.2 metres per second	15 metres per second	25 metres per second

Question 9

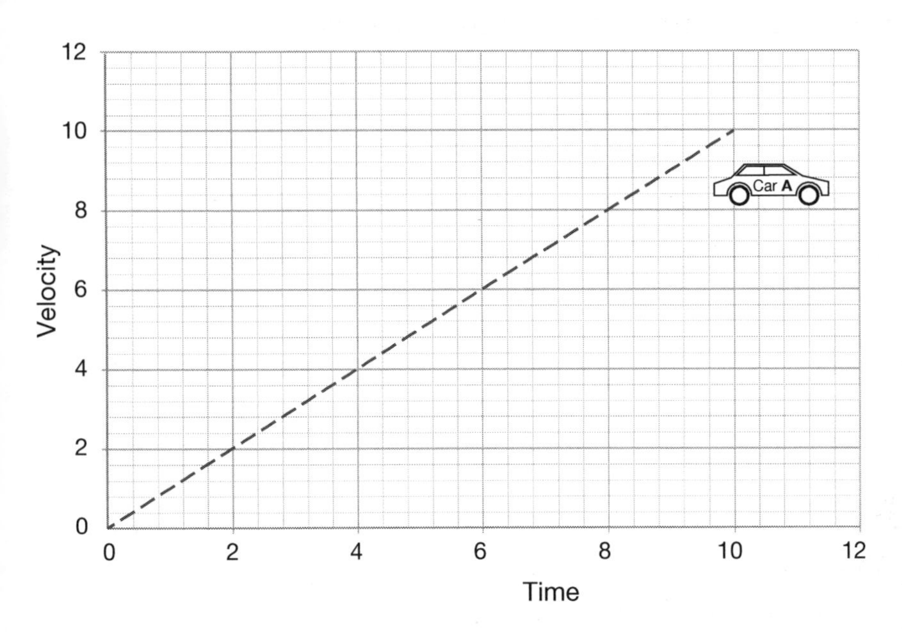

A car starts from rest, and within 10 seconds, it is travelling at 10 m/s.

What is the acceleration?

Answer

Question 10

A square field, S, has an area greater than 6400m2. Its length is increased by 31m and its width is also increased by 35m to give a rectangular field, R. Which one out of the following is true?

A. Area S > area R and perimeter S > perimeter R
B. Area S = area R and perimeter S = perimeter R
C. Area S < area R and perimeter S < perimeter R
D. Area S < area R and perimeter S > perimeter R
E. Area S > area R and perimeter S = perimeter R

Answer

Question 11

Points A and B are shown on the centimetre grid below (not to scale).

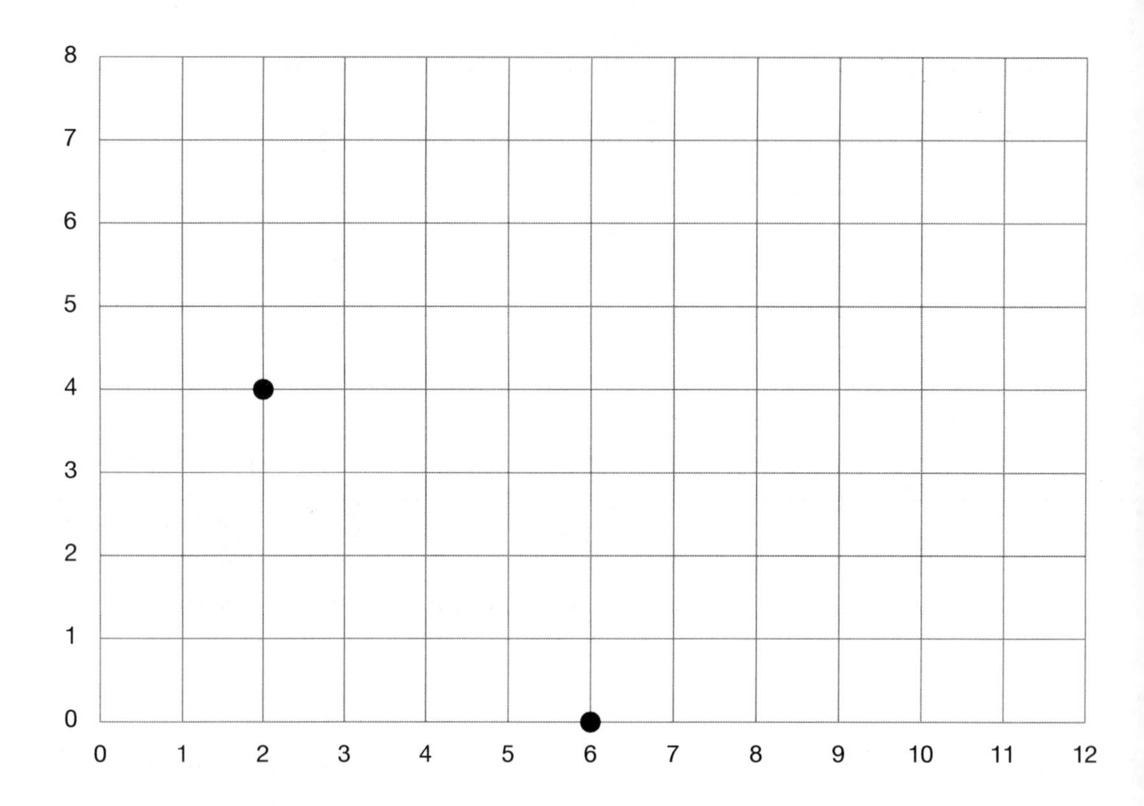

If you were to put another point at the co-ordinates (10, 4), what is the area of the triangle made by each of the three points?

Answer

Question 12

A bank pays 6.8% **compound** interest per year on an investment of £7000.

What is the value of the investment after two years?

Answer

Question 13

A train travelled 240 kilometres (km) at a speed of 80 kilometres per hour (kph).

How long did the journey take?

Answer

Question 14

Below is a table listing the percentage changes from 2012 to 2014 for five different companies.

Company	Percentage Change from 2012-2013	Percentage Change from 2013-2014
Company A	+17%	-5%
Company B	+12%	+5%
Company C	-11%	+8%
Company D	-5%	-7%
Company E	+8%	-3%

At Company E, the amount of sales for 2013, was what percent of the amount of sales for 2014? To one decimal point.

A	B	C	D
3.1%	97.3%	103.1%	115.9%

Question 15

Below is a table listing the percentage changes from 2012 to 2014 for five different companies.

Company	Percentage Change from 2012-2013	Percentage Change from 2013-2014
Company A	+17%	-5%
Company B	+12%	+5%
Company C	-11%	+8%
Company D	-5%	-7%
Company E	+8%	-3%

Using the above table, if company B earned £412,500 in 2012, how much money did the company make in 2014?

A	B	C	D
£316,875	£462,00	£415,290	£485,100

Question 16

Study the following graph carefully and answer the questions given below.

Distribution of candidates who were enrolled for a fitness course and the candidates (out of those enrolled) who passed the course in different institutes.

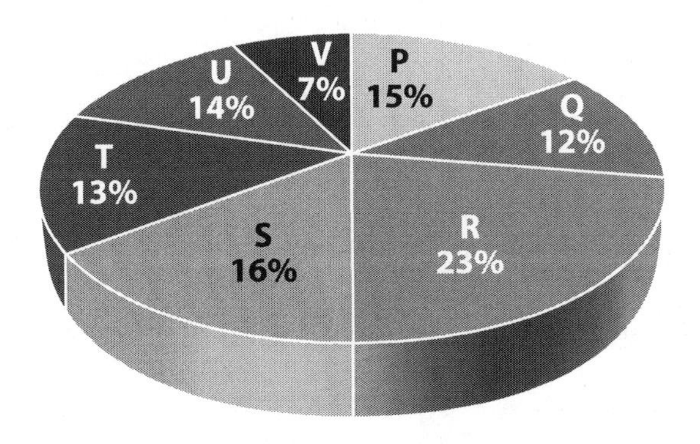

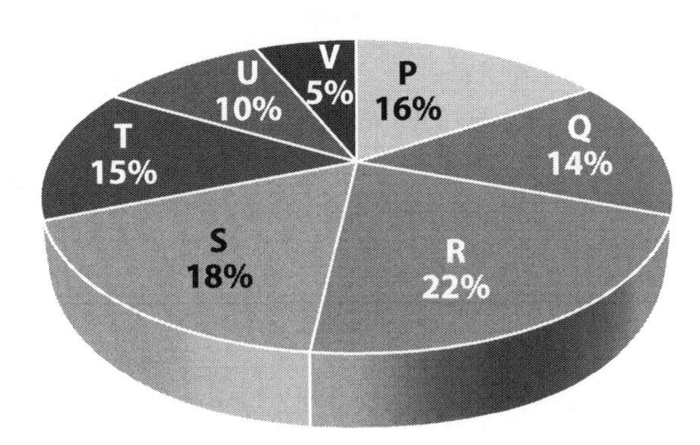

Which institute has the highest percentage of candidates passing the selection process to candidates enrolled?

A	B	C	D
Institute P	Institute Q	Institute T	Institute V

Question 17

The table shows the total tax paid in $ on annual taxable income.

Annual Taxable Income ($)	Tax Rate	Total Tax paid ($) at the top of this taxable income bracket
0-8,950	10%	895
8,951-36,250	15%	5,438
36,251-87,850	25%	21,963
87,851-183,250	28%	51,310
183,251-400,000	33%	132,000
Over 400,000	39.6%	

For example, a person with an annual taxable income of $60,000 will pay $5,438 plus 25% of ($60,000 - $36,251).

Tom has an annual taxable income of $22,500. The income tax, to the nearest $, he has to pay is:

A	B	C	D
$1,895.35	$2,175.35	$2,927.35	$22,85.35

Question 18

A travel company sells 3080 UK holidays in 2014. It expects the number it sells to increase by 14% each year. Work out the number of UK holidays the company expects to sell in 2019. Round all numbers up to the nearest whole number.

Answer [ ]

Question 19

2014 Sweet Sales in Supermarkets by region (100,000s)

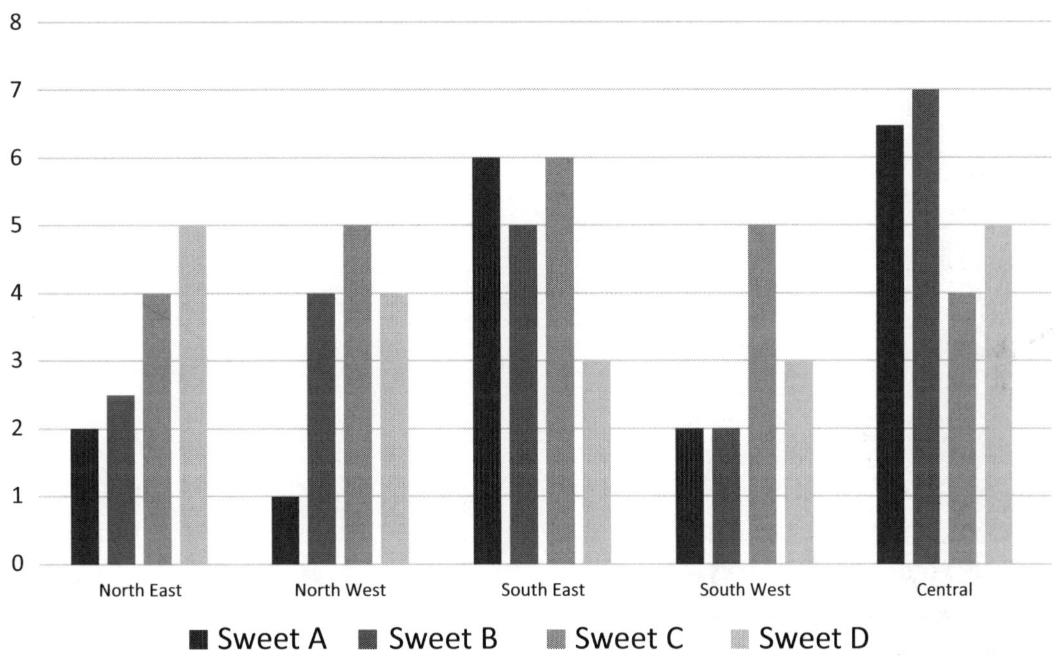

What was the total percentage increase for Sweet A's and C combined when comparing North East and South East regional supermarket sales?

A	B	C	D
150%	50%	250%	200%

Question 20

Using the above bar chart, what percentage of total sweet sales from all regions combined did the South East and Central region approximately contain? Rounded up to 1 decimal point.

A	B	C	D
51.9%	55.6%	62.8%	51.8%

Question 21

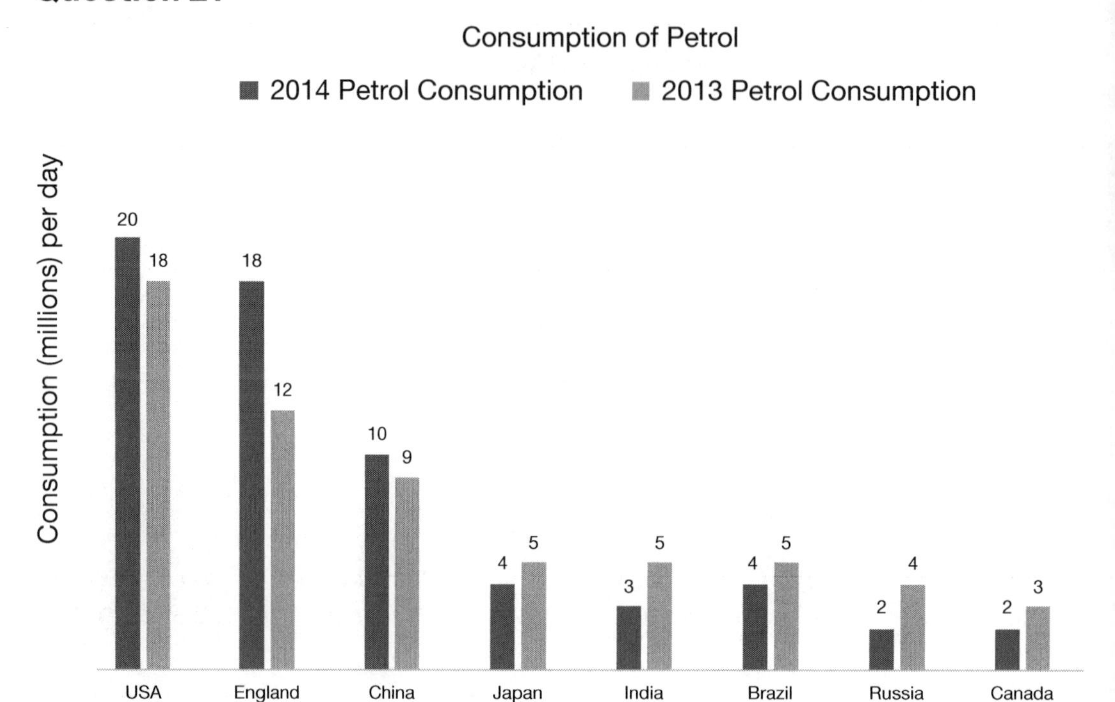

In England, if the petrol consumption per day continued to rise by 6.8% until 2016 and then decreased by 4% from 2016 to 2018, what would be the petrol consumption per day in 2018?

A	B	C	D
18.9 million	19 million	21.5 million	19.8 million

Question 22

Using the above bar chart, and assuming that China's percentage decrease in petrol consumption per day remains the same as it did between 2013 and 2014, what will be the estimated petrol consumption per day for China in 2017?

A	B	C	D
45,000	90,000	900,000	9,000

Question 23

Metal Orders for last year for Company A

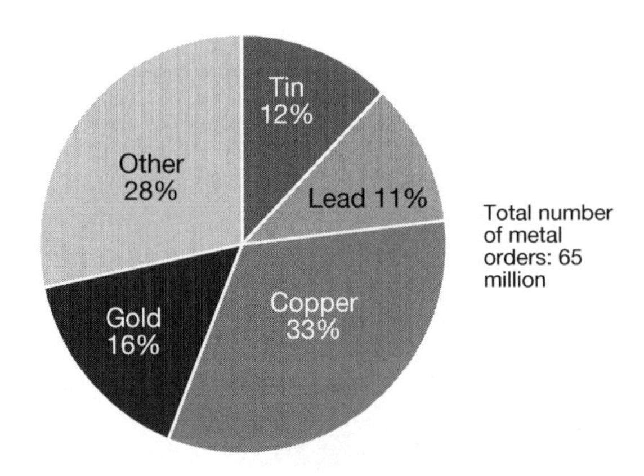

If the total income generated from gold for Company A was £310 million, approximately how much income would be generated per order of gold? Rounded up to two decimal points.

A	B	C	D
£29.82	£29.81	£29.80	£29.79

Question 24

Using the above pie chart, how many more copper orders took place compared to the number of orders of tin?

A	B	C	D
13,650,000	14,450,000	13,000,000	13,395,450

Question 25

Use the above pie chart to answer the question. If in the following year the number of orders of gold increases by 9.8%, and lead decreases by 3.4%, what is the ratio (in its simplest form) of gold to lead?

A	B	C	D
1464 : 1771	1771 : 2928	2928 : 1771	1348 : 1772

Question 26

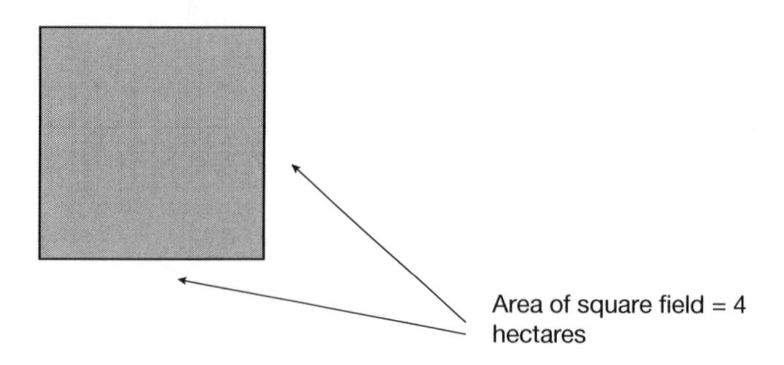

If the area of a square field is 4 hectares, then its side is:

A	B	C	D
200 cm	200 hectares	200 m	20 m

Question 27

50 men take a dip in a swimming pool 40m long and 20m broad. If the average displacement of water by a man is $4m^3$, then the rise in the water level in the tank will be?

A	B	C	D
50 cm	25 cm	200 cm	2 metres

Question 28

How many bricks, each measuring 22 cm x 11cm x 8 cm, will be needed to build a wall of 8m x 6m x 25 cm?

A	B	C	D
6135	6192	6198	9817

Question 29

Below is a diagram showing the plan of a building site.

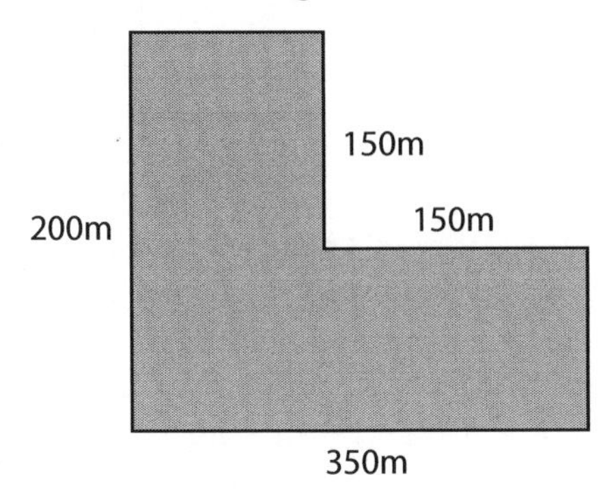

What is the area of the construction site? Give your answer in hectares.

1 hectare = 10,000 m = 2.47 acres.

A	B	C	D
475 hectares	47.5 hectares	40.75 hectares	4.75 hectares

Question 30

The following table shows the cost of booking holidays from a travel agent for next year.

HOLIDAY PRICES				
Types of Holiday Deals	**Turkey**	**Mexico**	**America**	**Spain**
All inclusive	£276pp	£720pp	£880pp	£320pp
Half board	£220pp	£640pp	£795pp	£275pp
Self-Catering	£180pp	£550pp	£620pp	£235pp

Work out the difference in cost of booking three all-inclusive holidays to Mexico, for two people, instead of booking one-self-catering holiday to Turkey for five people?

A	B	C	D
£1,250	£3,420	£9,000	£4,500

ANSWERS TO NUMERICAL REASONING – ADVANCED (Section 2)

Q1. C = 6 years

EXPLANATION = the annual population growth will remain the same.

Step 1 = Add 120,600 to 10 million (this is what you need to end up with).

120,600 + 10,000,000 = 10,120,600

Step 2 = starting with 10,000,000 and increasing the population by 1.002 (it is being increased each year, so 0.2 will become 0.2 ÷ 100 = 0.002 = increasing by 100% so it becomes = 1.002.

Step 3 = Population after year 1 = 1.002 x 10,000,000 = 10,020,000.

Population after year 2 = 1.002 x 10,020,000 = 10,040,040.

Population after year 3 = 1.002 x 10,040,040 = 10,060,120.

Population after year 4 = 1.002 x 10,060,120 = 10,080,240

Population after year 5 = 1.002 x 10,080,240 = 10,100,400.

Population after year 6 = 1.002 x 10,100,400 = 10,120,600.

Step 4 = we can stop here because we have reached the sum that was calculated in Step 1. So, it will take 6 years for the population of Country B to become 10,120,600 which is an increase in population of 120,600.

Q2. B = Country B

EXPLANTION = the annual population growth rate and unemployment rate remains the same from 2012 to 2014.

Step 1 = you need to work out the population for 2014 for each country.

Country A = 6,000,000 x 1.005 x 1.005= 6,060,150.

Country B = 10,000,000 x 1.002 x 1.002 = 10,040,040.

Country C = 10,000,000 x 1.008 x 1.008 = 10,160,640.

Country D = 12,000,000 x 1.008 x 1.008 = 12,192,768.

Step 2 = you then need to multiply these new populations by the unemployment rate.

Country A = 6,060,150 ÷ 100 x 8.5 = 515,112.75.

Country B = 10,040,040 ÷ 100 x 8.2 = 823,283.28.

Country C = 10,160,640 ÷ 100 x 7.4 = 751,887.36.

Country D = 12,192,768 ÷ 100 x 6.5 = 780,000.

Step 3 = So, Country B has the greatest unemployment rate in 2014.

Q3. D = ¼

EXPLANATION = number of bottles with TDS greater than 50 = 85 + 65 + 50 = 200.

Number of bottles with TDS 4 = 50. So 50/200 = 25/100 = 5/20 = 1/4

Q4. A = £172.41

EXPLANATION = this question may seem difficult because the table does not mention anything in relation to pounds sterling. However, you can use the sell at rate and divide it by the amount (200 euros).

So, 200 ÷ 1.16 = 172.413. To 1 decimal place = £172.40.

Q5. A = 150,000,000

EXPLANATION = 1.5 x 10^{11} (10 x 10 x 10 x 10 x 10 x 10 x 10 x 10 x 10 x 10 x 10) = 150000000000 ÷ 1,000 = 150000000 km.

Q6. C = 2 per cent

EXPLANATION = 4090 – 3040 = 1050.

So, 1050 ÷ 4090 x 100 = 25.67.

25.67 ÷ 15 (number of years between 1960 and 1975) = 1.71. Round up to nearest whole number = 2%.

Q7. 15 metres

EXPLANATION = in the first 3 seconds, the distance the cheetah runs is 45 metres. To work out the average: 45 ÷ 3 = 15 metres.

Q8. B = 10.2 metres per second

EXPLANATION = 102 ÷ 10 = 10.2 metres per second.

Q9. 1 m/s

EXPLANATION = change in velocity (10 – 0 = 10) ÷ time taken = 10 seconds.

So, 10 ÷ 10 = 1.

Q10. C = Area S < area R and perimeter S < perimeter R

EXPLANATION = if the perimeter is increased on both sides of the Square field S, that means the area of square field R is going to be bigger. This is also true about the perimeter; if both sides are increased in size to form field R, which means the perimeter for R is going to be bigger than that of perimeter S. So, the correct way to demonstrate this is answer C.

Q11. 16 cm^2

EXPLANATION = ½ x base x height. So, ½ x 8 x 4 = 16 cm^2.

Q12. £7984.37

EXPLANATION = for this question, it is vitally important to remember that interest will be added on to previous interest.

Step 1 = for the first year = 7,000 ÷ 100 x 6.8 = £476.

So, 7,000 + 476 = 7,476.

Step 2 = for the second year = 7,476 ÷ 100 x 6.8 = 508.37.

So, 7,476 + 508.37 = £7984.37.

Q13. 3 hours

EXPLANATION = time = speed ÷ change in velocity.

So, 240 ÷ 80 = 3 hours.

Q14. C = 103.1%

EXPLANATION = If A is the dollar amount of sales at Company E for 2013, then 3 percent of A, or 0.03 , A is the amount of decrease from 2013 to 2014. Thus 0.03 = 0.97 (to make a whole one).

Therefore, the desired percent can be obtained by dividing A by 0.97. So, 1 ÷ 0.97 = 1.0309 x 100 = 103.09. Expressed as a percentage to the nearest tenth = 103.1%.

Q15. D = £485,100

EXPLANATION = According to the table, if the dollar amount of sales at Company B was £412,500 for 2012, then it was 12 percent greater for 2013, which is 112 percent of that amount. So, 412,500 ÷ 100 x 112 = 462,000. From 2013 to 2014, the company saw a 5% increase, which is 105% of the previous month. So, 462,000 ÷ 100 x 105 = 485,100. So, the correct answer is £485,100.

Q16. B = Institute Q

EXPLANATION =

$$P = \left[\left(\frac{16\% \text{ of } 920}{15\% \text{ of } 1500}\right) \times 100\right]\% = \left[\frac{16 \times 920}{15 \times 1500} \times 100\right]\% = 65.42\%.$$

$$Q = \left[\left(\frac{14\% \text{ of } 920}{12\% \text{ of } 1500}\right) \times 100\right]\% = 71.56\%.$$

$$R = \left[\left(\frac{22\% \text{ of } 920}{23\% \text{ of } 1500}\right) \times 100\right]\% = 58.67\%.$$

$$S = \left[\left(\frac{18\% \text{ of } 920}{16\% \text{ of } 1500}\right) \times 100\right]\% = 69\%.$$

$$T = \left[\left(\frac{15\% \text{ of } 920}{13\% \text{ of } 1500}\right) \times 100\right]\% = 70.77\%.$$

$$U = \left[\left(\frac{10\% \text{ of } 920}{14\% \text{ of } 1500}\right) \times 100\right]\% = 43.81\%.$$

$$V = \left[\left(\frac{5\% \text{ of } 920}{7\% \text{ of } 1500}\right) \times 100\right]\% = 43.81\%.$$

So, the institute with the highest percentage rate of candidates passed, to candidates enrolled, is Institute Q.

Q17. C = $2927

EXPLANATION = 22,500 – 8,951 = 13,549.

15% of 13,549 = 13,549 ÷ 100 x 15 = 2032.35.

Add the 10% of the previous sum (8,951) = 895.

So, 895 + 2032.35 = 2927.35. To the nearest $ = $2,927.

Q18. 5929

EXPLANATION = to work out the percentage increase from 2014 to 2019, you need to work out the total number of holidays for each year, using the 14% increase each year.

So, in 2014 = 3080 holidays.

In 2015 = 3080 ÷ 100 x 114 = 3511.

In 2016 = 3511 ÷ 100 x 114 = 4002.

In 2017 = 4002 ÷ 100 x 114 = 4562.

In 2018 = 4562 ÷ 100 x 114 = 5201.

In 2019 = 5201 ÷ 100 x 114 = 5929.

Q19. C = 250%

EXPLANATION = First, let's work out the percentage increase for sweet A.

North East = 200,000, South East = 600,000. To work out the percentage increase we use this formula: % increase = (highest number – lowest number) ÷ lowest number x 100.

So, 600,000 – 200,000 = 400,000 ÷ 200,000 = 2 x 100 = 200%.

For Sweet C, we use the same method. North East = 400,000, South East = 600,000.

So, 600,000 – 400,000 = 200,000 ÷ 400,000 = 0.5 x 100 = 50%.

So, the total percentage increase for both sweets, in both regions = 200 + 50 = 250%.

Q20. D = 51.8%

EXPLANATION = the first step is to work out what the total number of sweet sales were: North East = 13.5, North West = 14, South East = 20, South West = 12, Central = 22.5.

So, 13.5 + 14 + 20 + 12 + 22 = 82. So technically, in terms of (100,000's) = 8,200,000.

Next, you will need to know the sweet sales for South East and Central regions = 20 + 22.5 = 42.5 x 100,000 = 4,250,000.

So to work out what percentage of the total sales are accounted for by South East and Central regions = 4,250,000 (accounted regions) ÷ 8,200,000 (total sales) x 100% = 51.82%

Q21. A = 18.9 million

EXPLANATION = first, you need to work out the percentage increase each year from 2014 to 2016.

So, in 2014 there is 18 (million); to work out a 6.8% increase would equal 106.8%. So, 18 ÷ 100 x 106.8 = 19.2 (million). This is the consumption for 2015. From 2015 to 2016, the same thing applies. 19.2 ÷ 100 x 106.8% = 20.5 (million).

From 2016 to 2017, there is a 4% decrease. So, 20.5 ÷ 100 x 96% = 19.7 (million). From 2017 to 2018 = 19.7 ÷ 100 x 96% = 18.9 (million).

Q22. D= 9,000

EXPLANATION = first, you need to work out what the percentage decrease from 2013 to 2014 was. Use the formula: percentage decrease = highest number – lowest number ÷ highest number x 100.

So, 10 – 9 = 1 ÷ 10 x 100 = 10%. So, the percent decrease which remains constant is 10%.

To work out 2015 = 9 (million) ÷ 10 = 900,000.

To work out 2016 = 900,000 ÷ 10 = 90,000.

To work out 2017 = 90,000 ÷ 10 = 9,000.

Q23. B = £29.81

EXPLANATION = first, to work out the approximate income per order of gold, you need to work out what percentage of gold of the total number of orders.

So, 65,000,000 ÷ 100 x 16% (gold's percentage of total) = 10,400,000.

You are told that £310 million was the total generated income for the gold orders. So, you need to work out the value per order of gold metal.

310,000,000 ÷ 10,400,000 = 29.807.

Rounded up to two decimal points and the value per order of gold = £29.81.

Q24. A = 13,650,000

EXPLANATION = to work out the number of orders for tin = 65,000,000 ÷ 100 x 12 = 7,800,000.

To work out the number of orders for copper = 65,000,000 ÷ 100 x 33 = 21,450,000.

So, the difference between the number of order of tin compared to the number of orders of copper = 21,450,000 – 7,800,000 = 13,650,000.

Q25. C = 2928 : 1771

EXPLANATION = first you need to work out the total for gold.

So, 65,000,000 ÷ 100 x 16 = 10,400,000. This increases by 9.8%, which is 109.8%. So, 10,400,000 ÷ 100 x 109.8 = 11,419,200.

Then work out the total for lead.

So, 65,000,000 ÷ 100 x 11 = 7,150,000. This decreases by 3.4%, which is 96,6%. So, 7,150,000 ÷ 100 x 96.6 = 6,906,900.

So, the ratio is 11,419,200 : 6,906,900. This needs to be simplified into its simplest form.

Both sides of the ratio can be divided by 3,900 which gives = 2928 : 1771. This is its simplest form because no other numbers go into both 2928 and 1771.

Q26. C = 200m

EXPLANATION = 4 hectares needs to be converted into metres. So 4 hectares x 10,000 (1 hectare = 10,000m) – 40,000 m^2.

The square root of 40,000 (in order to work out both sides of the square field) = 200m. (40,000 ÷ 2 = 20,000, however to work out the sides of each square, take off a zero for each side, so you need to take off two zeros, to work out the area = 200 x 200 = 40,000.

So, the correct answer = 200m.

Q27. B = 25 cm

EXPLANATION = Total volume of water displaced = $4m^3$ x 50 (men) = $200m^3$.

Rise in water level = 200 ÷ (40 x 20) = 200 ÷ 800 = 0.25m x 100 = 25cm.

Q28. C = 6198

EXPLANATION = Number of bricks = Volume of the wall ÷ volume of 1 brick = 800 x 600 x 25 = 12,000,000.

Volume of 1 brick = 22 x 11 x 8 = 1936.

So, 12,000,000 ÷ 1936 = 6198.

Q29. D = 4.75 hectares

EXPLANATION = 350 x 200 = 70,000.

150 x 150 = 22,500. So, 70,000 – 22,500 = 47,500

47,500m ÷ 10,000 = 4.75 hectares.

Q30. B = £3,420

EXPLANATION = Self-catering holiday to Turkey for 5 people = 180 x 5 = 900.

All-inclusive holiday to Mexico for 2 people = 720 x 2 = 1440. Booked three times = 1440 x 3 = 4320.

So, 4320 – 900 = 3,420.

A FEW ***FINAL WORDS...***

You have now reached the end of your Numerical Reasoning testing guide. Hopefully, you now feel more comfortable and confident with these tests, and will be able to successfully pass any Numerical Reasoning test.

For any psychometric test, there are a few things to remember to help you perform at your best...

REMEMBER – THE THREE P'S!

1. **Prepare.** This may seem relatively obvious, but you will be surprised how many people fail psychometric testing because they lacked knowledge and understanding of what to expect. Be sure to practice these tests prior to sitting a real test. Not only will you become familiar with the testing questions, it will also take off some of the pressure leading up to that all important test. Like anything, the more you practice, the more likely you are to succeed!
2. **Perseverance.** You are far more likely to succeed at something if you continuously set out to achieve it. There are times when everybody experiences setbacks, or finds obstacles in the way of their goals. The important thing to remember when this happens, is to use those setbacks and obstacles as a way of progressing. It is what you do with your past experiences that helps to determine your success in the future. If you fail at something, consider 'why' you have failed. This will allow you to improve and enhance your performance for the next time.
3. **Performance.** Your performance will determine whether or not you are likely to succeed. Attributes that are often associated with performance are self-belief, motivation and commitment. Self-belief is important for anything you do in your life. It allows you to recognise your own abilities and skills and believe that you can do well. Believing that you can do well is half the battle! Being fully motivated and committed is often difficult for some people, but we can assure you that, nothing is gained without hard work and determination. If you want to succeed, you will need to put in that extra time and hard work!

Good luck with your Numerical Reasoning tests. We wish you the best of luck with all your future endeavours!

The how2become team

The How2become team

how2become